AF513091

INTRODUCTION TO BIOSCIENCE

INTRODUCTION TO BIOSCIENCE

By

Dr. J.P. Sharma

M.Sc., Ph.D., FISST

Ex. H.O.D. Deptt. of Botany

Hindu College, Sonepat

(M.D. University, Rohtak, Haryana)

UNIVERSITY SCIENCE PRESS

(An Imprint of Laxmi Publications Pvt. Ltd.)

An ISO 9001:2008 Company

BENGALURU • CHENNAI • COCHIN • GUWAHATI • HYDERABAD

JALANDHAR • KOLKATA • LUCKNOW • MUMBAI • RANCHI • NEW DELHI

BOSTON (USA) • ACCRA (GHANA) • NAIROBI (KENYA)

INTRODUCTION TO BIOSCIENCE

Edited by : **Dr. Pragya Dixit**

Printed and bound in India
Typeset at Excellent Graphics, Delhi
First Edition : 2016
ISBN 978-93-83828-22-7

Branches

✆	Bengaluru	080-26 75 69 30	
✆	Chennai	044-24 34 47 26,	24 35 95 07
✆	Cochin	0484-237 70 04,	405 13 03
✆	Guwahati	0361-254 36 69,	251 38 81
✆	Hyderabad	040-27 55 53 83,	27 55 53 93
✆	Jalandhar	0181-222 12 72	
✆	Kolkata	033-22 27 43 84	
✆	Lucknow	0522-220 99 16	
✆	Mumbai	022-24 91 54 15,	24 92 78 69
✆	Ranchi	0651-220 44 64	

PUBLISHED IN INDIA BY

UNIVERSITY SCIENCE PRESS
(An Imprint of Laxmi Publications Pvt. Ltd.)

An ISO 9001:2008 Company
113, GOLDEN HOUSE, DARYAGANJ,
NEW DELHI - 110002, INDIA
Telephone : 91-11-4353 2500, 4353 2501
Fax : 91-11-2325 2572, 4353 2528
www.laxmipublications.com info@laxmipublications.com

C—10063/015/03
Printed at: N. K. Book Binder, Delhi.

CONTENTS

PREFACE

In recent years there have been many advances made in the field of bioscience. And so the techniques of bioscience have been included in the curriculum of Engineering sciences of Indian Universities. There has been made tremendous progress in the research methodology of biosciences. The progress has been driven by the new and minor modifications to existing techniques in turn propelling and enhancing the scientific investigation. However, the use of any procedure can only be advantageous if there is an appreciation of basic principles, requirements, benefits and pitfalls of the technique employed. This is true for both experienced and beginners in the field of biosciences.

No doubt, the newly enrolled students for this course may find struggling to understand the basics of techniques in bioscience as there is a lack of appropriate support in the literature on this topic. The book entitled 'Introduction to Bioscience' has been written keeping this in view. With this in mind, the rationale of this book is to supply the students with an easy to use reference book that will provide the students with a basis for understanding the basic bioscience techniques that the students come across in their studies. The book provides the foundation of biosciences techniques to make it a novice course book for the students as well as teachers. The content of the book follow the latest curriculum prescribed by the universities for B.Tech. courses.

All efforts have been made to make the subject matter comprehensive up to date and concise. Every care has been taken to make the language simple and the presentation of the text systematic. Well informative and self explanatory diagrams have also been included in sufficient number. While great emphasis has been laid on basic fundamentals, no effort has been spared to incorporate recent information in sufficient details. Each chapter has been appended with sufficient number of review exercises that include multiple choice type questions, objective type questions, and descriptive questions.

It is confidently hoped that the present book will go a longway in meeting the requirements of students who aspire for brighter career.

Constructive suggestions for the improvement of book shall be gratefully acknowledged.

— Author

UNIT–1

INTRODUCTION TO BIOSCIENCE

1 Introduction To Bioscience

- Defining Characteristics of Living Beings
- Comparison in Structure and Organisation between Living and Non-living Things
- Levels of Organisation
- Review Exercises
- Test Questions

The living world comprises of diverse type of organisms. Millions types of living organisms exist on the Earth. *The Science which deals with the study of living things is called biology* (Gk. *bio*-life, *logos*-study or discourse). The definition of biology appears to be quite simple, but this science is very complex and informative. It covers all aspects of study of living beings like origin, occurrence, external form, organisation, life history and inheritance. All these disciples constitute several aspects or branches of biology. Being broad based and multidisciplinary, the term biology is often replaced by the term **'life sciences'** or **biosciences** or **biological sciences.**

Biology deals with the science of living things–what they are, how they work, interact and change or evolve. The living organism interact with one another, as well as with their physical and chemical environment. Biology as a disciple of science, demands a scientific approach for analysis of its facts. It is therefore, imparative that student of biology should have a clear idea about the basic nature of science.

There are two types of objects—living and non-living. However, despite our recognition of the two, it is very difficult to distinguish living beings from non-living objects. Though biologists have vast knowledge of living things, yet it is quite difficult for them to definite life. Therefore, rather than to trying to definite life, biologists focus on how life works. All the living beings share certain unified and basic characteristics.

Life is a unique, complex organisation of molecules expressing itself through chemical reactions, which lead to growth, development, responsiveness, adaptation and reproduction. The object by itself exhibiting the growth, development, death etc., is designated as living being.

DEFINING CHARACTERISTICS OF LIVING BEINGS

Although, there exist a great diversity among organisms, still they have many things in common. The diverse form of organisms such as fungi, plants, insects and vertebrates are made up of cells, similar in internal organisation and functions. May be a single celled *Amoeba* or bacterium or giant multicellular whale or the unique human, all share some unified and basic characteristics like growth, reproduction, ability to sense environment, metabolism, ability to self replicate, self organise, interact and emergence, etc.

All these become the defining property of living organisms. Let us understand each of these.

1. **Growth.** Increase in mass and increase in number of individuals are the two characteristics of growth. All organisms are able to grow. A multicellular organism grow by cell division. In plants growth occurs throughout their lifespan. However, in animals this growth occurs only upto a certain age, and thereafter cell division occurs in certain tissues to replace the lost cells. Unicellular organisms also grow by cell division. The increase in body mass is considered as growth. In living organisms growth is from inside.

Certain non-living objects such as mountains, boulders and sand mounds also grow. However, the growth of non-living objects is established by the accumulation of material on the surface. Growth, therefore, cannot be taken as a defining characteristic of living systems.

2. **Reproduction.** Living organisms produce new individuals similar to them. Organisms reproduce by sexual and asexual means. Fungi multiply and spread easily by producing millions of asexual spores. Some organisms like yeast and *Hydra* multiply by budding. The flat worms (*e.g., Planaria*) regenerates from fragmented parts of their body. The filamentous algae, fungi, protonema of mosses also multiply by fragmentation.

In unicellular organisms like bacteria, *Amoeba,* unicellular algae reproduction is synonymous with growth *i.e.,* increase in number of cells. Therefore, in unicellular organisms, there is no distinction in the usage of the terms 'growth' and 'reproduction'.

Many organisms such as mules, sterile worker bees, infertile human couples, etc., do not reproduce. Also non-living objects cannot reproduce and replicate by itself. Therefore, reproduction cannot be taken as a defining property of living organisms.

3. **Metabolism.** Living organisms are made of different kinds of chemicals, which are constantly being made and changed into some other biomolecules. These reactions are called **metabolic reactions.** Thousands of metabolic reactions occur simultaneously inside all living organisms. The sum total of all the chemical reactions occurring in our body is called **metabolism.**

Metabolic reactions can be performed outside the body is cell free systems. An isolated metabolic reaction outside the body of an organism is neither living nor non-living. The isolated metabolic reactions *in vitro* are not 'living things' but are 'living reactions'. Hence, metabolism can be considered defining feature of all living organisms without exception.

4. **Cellular Organisation.** All types of organisms starting from simple to highly complex entities are made up of one or more cells, similar in internal structure and functions. Hence, cellular organisation of the body is the defining feature of life forms.

5. **Protoplasm.** Living beings possess a highly organised living matter, called **protoplasm (Purkinji,** 1837) capable of performing life processes. **Huxley** (1868) has described it as **'the physical basis of life'.** The protoplasm or living matter consists of a variety of organic and inorganic substances. Non-living things lack such living matter.

6. **Consciousness.** All living organisms have the ability to sense their surroundings or environment and respond to these environmental stimuli, which could be physical, chemical or biological. All organisms, from prokaryotes to most complex eukaryotes can sense and respond to environmental cues. Photoperiod (*i.e.,* duration of light) affects reproduction in seasonal breeders of both plants and animals. Plants respond to external factor like light, water, temperature, pollutants and other organisms. All organisms are able to handle chemicals that enters their bodies, and are aware of their surroundigns. We sense our environment through our sense organs. Human being is the only organism that has 'self consciousness' *i.e.,* aware of himself. Hence, consciousness (chetan tatwa) is the defining property of living organisms.

It is quite difficult to define the living state of a human being at different times. For example, lying in coma in hospital, supported by machines (*i.e.*, heart and lung machines), with dead brain. The patient has no self consciousness. Are such patients who never come back to normal life, living or non-living.

All living phenomenon are due to underlying interactions. The properties of tissues are not present in constituent cells but arise as a result of interactions among the constituent cells. In the same way, the properties of tissues are not present in the molecular constituents of the organelle, but is due to the interactions among the molecular components that constitute the organelle. These interactions result in emergent properties at a higher level of organisation. Thus, *the living organisms are self-replicating, evolving and self regulating interactive systems capable of responding to external stimuli.* All living organisms—present, past and future, are linked to one another by the sharing of common genetic material, but to varying degree.

COMPARISON IN STRUCTURE AND ORGANISATION BETWEEN LIVING AND NON-LIVING THINGS

Living beings share some features with the non-living things whereas they differ in certain other respects. Some of these points of similarities and differences are listed below :

Similarities

- Living beings are made up of same elements, which constitute the non-living things.
- These elements combine and form **molecules (biomolecules** in living beings).
- The molecules of living beings or biomolecules are capable of interacting like those of non-living things.
- Both the living beings and non-living things are governed by the same physical laws such as those of gravitation, magnetism, action and reaction, etc.

Differences

- The differences between living beings and non-living things at molecular level started very early in the geographical history when organic compounds were synthesized for the first time.
- Living beings possess a number of organic molecules, whereas the non-living things do not have them.
- Living beings possess a number of complex molecules called **macromolecules** formed by the condensation and interaction of smaller molecules known as **micromolecules.** The macromolecules are absent in non-living things.
- Living beings possess DNA or RNA as genetic material. Such genetic material is not present in non-living things.
- In the living beings, the molecules and compounds coordinate with one another to form sub-cellular components such as **cell organelles,** cytoplasmic matrix etc. On the other hand in non-living things, molecules and compounds are organised into **colloids, crystals** and **mixtures** and there is no further organisation beyond these structures.
- In the living word, the sub-cellular components are organised into the protoplasmic units called **cells.** The cells are organised into **tissues,** tissues into **organs,** organs

into **organ systems** and the organ systems into an **individual.** No such type of organisation is found in non-living things.

- In living beings, organisation becomes more and more complex, beyond the individual level.

LEVELS OF ORGANISATION

The term organisation has many meanings depending on the context in which it is used. In a simple way organisation may be defined as *'the manner in which smaller components of any structure, system or a situation are arranged in a hierarchy to co-ordinate with one another towards a goal'.*

The above definition indicates the presence of organisation at several levels. The levels of organisation of matter in the living and non-living things are different from one another (Fig. 1.1 and Fig. 1.2). The living beings present a characteristic complexity in their organisation. Various levels of organisation in the living beings and the non-living things are described below :

(*i*) **Atomic Level.** The basic unit or component of organisation of both living and non-living objects is the **atom.** The atoms are, of course, themselves made of still smaller particles—mainly **electrons, protons** and **neutrons.**

(*ii*) **Molecular Level.** Atoms are organised to form molecules and compounds. In non-living world, the molecules and compounds are aggregated to form **crystals, mixtures** and **colloids,** but beyond them the organisation is absent. In the living world, the molecules and compounds co-ordinate with one another to form sub-cellular components like cell organelles, cytoplasmic matrix, etc.

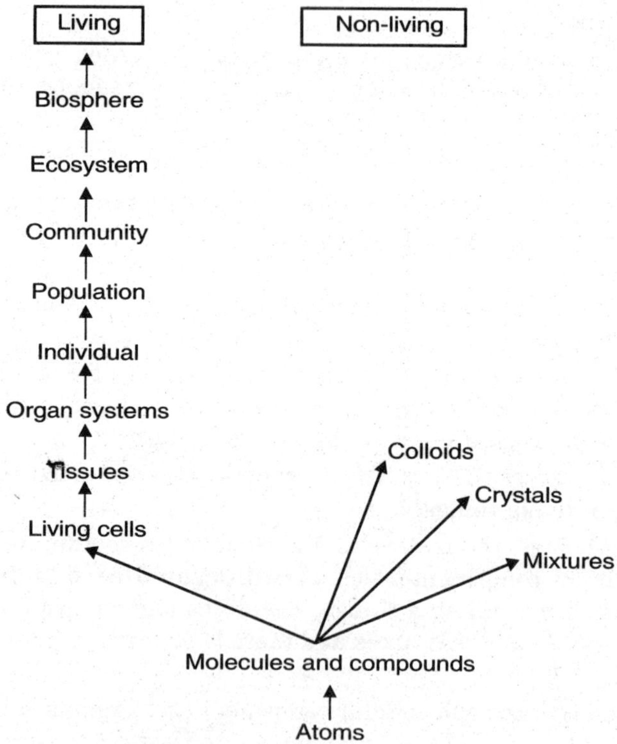

Fig. 1.1. Levels of organisation in living and non-living.

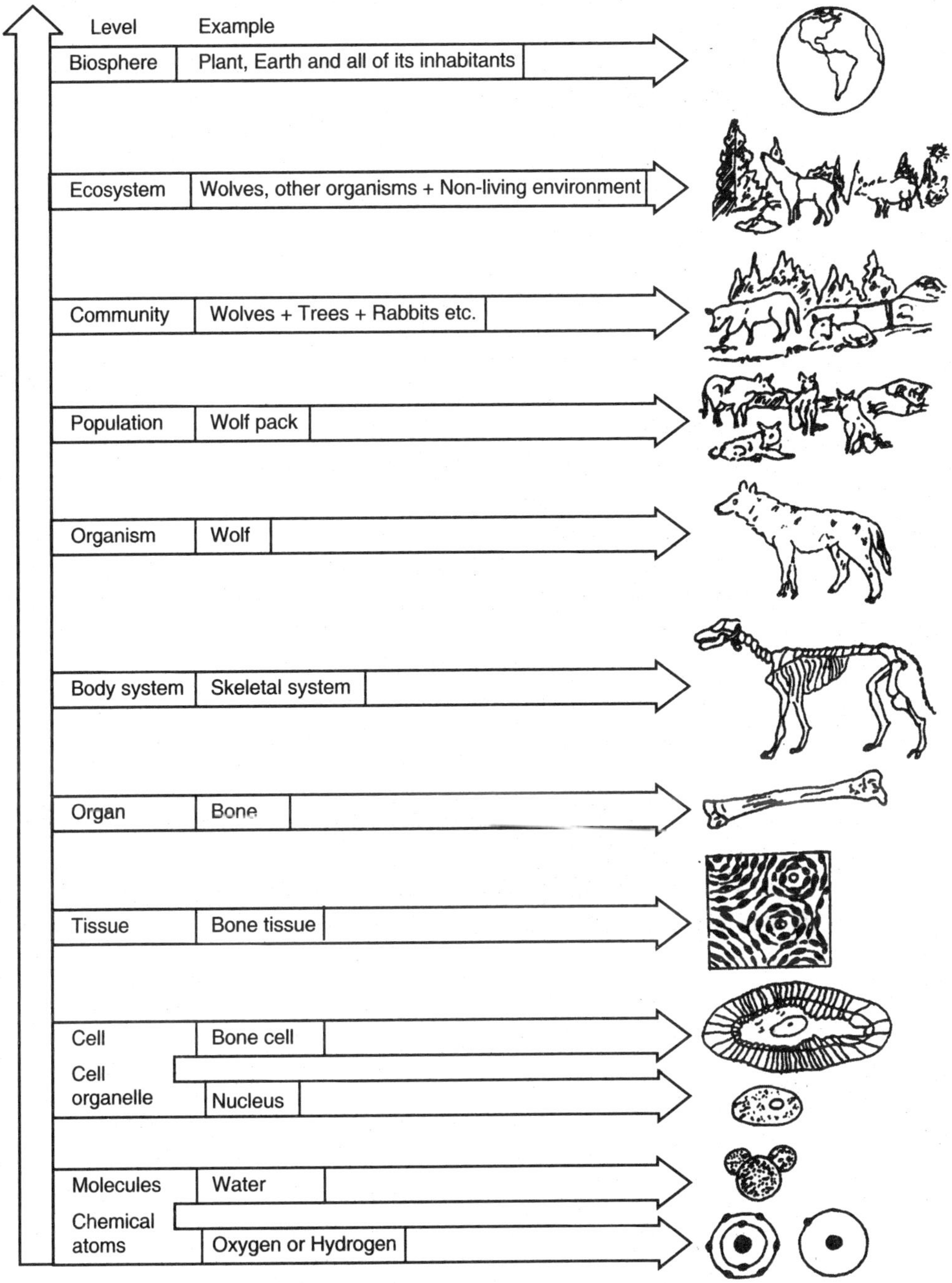

Fig. 1.2. Levels of biological organization.

(iii) **Cellular Level.** The sub-cellular components co-ordinate their activities to produce living cells. The cells are the structural and functional units of life. Living beings are made up of one cell **(unicellular organisms)** or numerous cells **(multicellular organisms).**

(iv) **Tissue Level.** In simple multicellular organisms similar cells form aggregates. This type of organisation is called **colonial.** In most multicellular organisms cells are organised in

more efficient manner to form **tissues.** *'A tissue is a group of similar cells, which are specialized to perform a specific function'.*

(*v*) **Organ Level.** Various tissues get associated to form a next complex organisation, called **organ.** An organ is much more efficient structure than any single tissue to perform a particular function.

(*vi*) **Organ System Level.** In complex organisms, several organs function together in a group called **organ system.** Different organs in an organ system work like a team and accomplish more complex type of function in the body.

(*vii*) **Individual Level (Organism Level).** Different organ systems function together to form an integrated unit called **individual** or **organism.** It is a self sufficient unit of the living world.

(*viii*) **Population Level.** Individuals of the same species living at one place, where they can interact with each other constitute a **population.**

(*ix*) **Community Level.** Different populations of plants and animals interact with one another to form a still higher organisation called **community.**

(*x*) **Ecosystem Level.** The living organisms cannot survive without abiotic environment, which supplies necessary raw materials to biotic-communities. The communities and their abiotic environment together constitute an **ecosystem.**

(*xi*) **Biosphere Level.** It is the highest and most complex level of organisation of the living world. All the ecosystems of the world interact with one another to constitute the **biosphere** or **ecosphere.** It includes entire livable space of the earth.

An organism is, therefore, not only an individual but also a part of population, community, ecosystem and biosphere.

REVIEW EXERCISES

MULTIPLE CHOICE TYPE QUESTIONS

1. Which one of the following is a defining property of living beings.

(*a*) Growth (*b*) Reproduction
(*c*) Metabolism (*d*) Steady state.

2. The physical basis of life is

(*a*) Protoplasm (*b*) Cell
(*c*) Cell organelles (*d*) Biomolecules.

3. The differences between non-living things and living beings started at the level of

(*a*) Atoms (*b*) Molecule
(*c*) Tissue (*d*) Organ.

OBJECTIVE TYPE QUESTIONS

1. Fill in the blanks with suitable words:

(*i*) Increase in .... and increase in number of individuals are the two characteristics of growth.

(*ii*) In unicellular organisms reproduction is synonymous with ....

(*iii*) The sum total of all the chemical reactions occurring in the body of a living organism is called ....

(*iv*) All living organisms have the ability to sense their ....

(*v*) Living beings possess .... or .... as genetic material.

2. Match the items of Column I with those in Column II.

Column I	Column II
(*i*) Protoplasm	(*a*) Cell division
(*ii*) DNA	(*b*) Replication
(*iii*) Growth	(*c*) Response to stimuli
(*iv*) Reproduction	(*d*) Living matter
(*v*) Consciousness	(*e*) Genetic material

TEST QUESTIONS

1. Define biology. Why is this term often replaced by biosciences?
2. List the similarities between non-living things and living organisms.
3. Why are growth and reproduction not considered as defining properties of living beings?
4. Describe any two characteristics of living beings that are considered defining property of living organisms.
5. Differentiate between living beings and non-living things.
6. Explain the levels of organisation of matter in the living and non-living things.

2 Cell Structure and Function

Organisms are made up of one or more cells. Many organisms consist of just one cell. They are called unicellular organisms *e.g., Amoeba, Euglena, Chlamydomanas*, etc. There are other organisms which are composed of many cells. These are called multicellular organisms. Higher animals and plants are the examples of multicellular organisms. The body of a 60 kg, adult human being has about 100 trillion (100×10^{12}) cells. A drop of blood contains several million cells. The large sized organisms do not have large sized cells, instead they possess higher number of cells.

The study of form, structure and composition of cells is called **cytology.** The branch of biology that deals with various aspects of structure, chemistry, development, genetics and functioning of cells is called **cell biology.** In cell biology, scientists study fundamental processes that are common to all cells. Hence, cell biology appears to be a unifying subject and studying cell biology in some sense the same as studying life.

DISCOVERY OF CELL

The cell is the basic unit of structure and function of living organisms. The study of cell, not only required the development and improvement of microscope but also a variety of other techniques like staining to identify parts and study of other functions. The study of cell has continued for more than the last three and a half centuries. Investigations are still continuing to add to our knowledge about the cell.

The term 'cell' was introduced by an English scientist **Robert Hooke** in his book *Micrographia* published in London in 1665. He examined thin slices of cork under his primitive microscope (Fig. 2.1*a*). He observed a honey comb like pattern in the slice of cork. This honey comb consisted of thick walled box like compartments which he called the cells. He did not know the significance of these structures and regarded them as passages for conducting fluids. In fact his 'cell' (L. *Cella*-hollow spaces or compartments) were the cell walls enclosing the vacant spaces left by dead cork cells (Fig. 2.1*b*). The term 'cell' is a misnomner as a living cell is neither hollow nor always covered by a wall. However, the term is still used because its simple and a suitable substitute has not been found.

Hooke studied a wide range of plants under the microscope and concluded that they had the same structure.

Cells were also observed prior to Hooke, by **Malphighi** (1661), who called them **saccules** and **utricles.**

Leeuwenhoek (1672) assembled a simple microscope and observed bacteria, protozoa, red blood corpuscles, etc.

The study of cell progressed rapidly in nineteenth century because of great improvement in the microscope.

Robert Brown (1831) discovered the **nucleus** in the cells. By this time it had also been found that the cells were surrounded by some sort of limiting structure, the **cell membrane.**

Living semi-fluid material of cells was discovered by **Dujardin** (1838). He called this material **sarcode.**

Purkinje (1839) renamed sarcode or jelly like substance of the cells as **protoplasm** (Gk. *proto*-first, *plasma*-substance).

M.J. Schleiden (1804–1881), a German botanist, discovered that all plants were composed of cells and a year later.

Theodore Schwann (1810–1882), a German zoologist, observed that all animals were also formed of cells. He noted that except for the cell wall the plant and animal cells were more or less similar, having a nucleus located in a jelly like substance. He defined the cell as 'a membrane bound nucleus containing structure'. He proposed **cell hypothesis** stating the *'the bodies of plants and animals are composed of cells'*.

Schleiden and Schwann exchanged their findings and applied the cell concept to all organisms. This laid the foundation of the cell theory.

Robert Hooke (1635–1703) developed a crude microscope, with which he studied the structure of cork. He took a piece of cork and prepared thin slice by means of a sharp pen knife. The surface of slice was smoothened and the slice was further thinly pared to remove an extremely thin piece. The cork slice was now placed on a black object plate, because it was itself a white body. A deep planoconcave lens was used for throwing light on the cork piece. The later was observed under the microscope. The piece of cork was found to have a honey comb structure with a number of box like compartments each having a pore and separated from others by diaphragms. He named the compartments as cells (L. *cella*-hollow spaces or compartments).

Robert Hooke described his observations in his book **'Micrographia',** or *'Some Physiological Descriptions of Minutae made by magnifying glasses with observations and enquiries there upon'* published in London in 1665. The chapter which laid the foundation of cell biology as we know today is *'Observe 'XVIII' of the Schematisme or Texture of Cork and of the Cells and Pores of some other such frothy bodies.'* Hooke made his discoveries inspite of primitive cutting instruments, dimlight sources and lenses with aberrations.

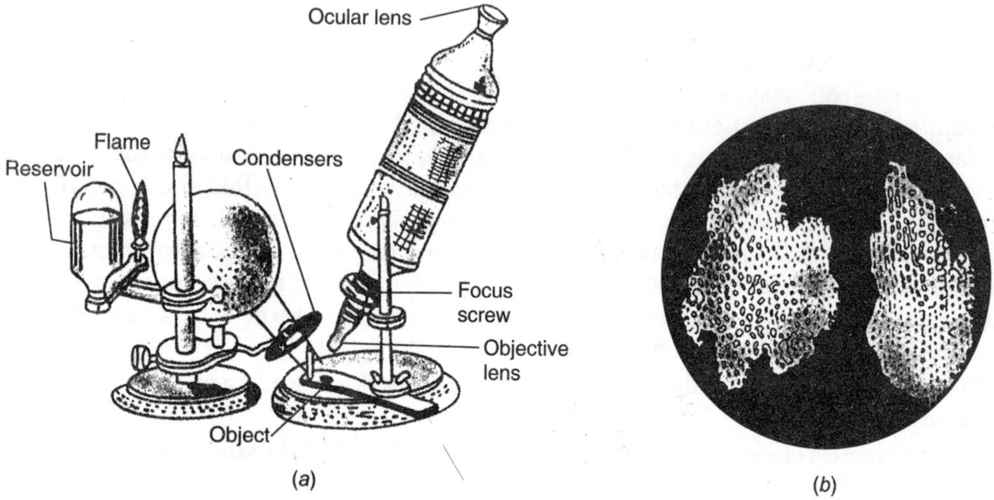

Fig. 2.1. (*a*) A crude microscope used by Robert Hooke and (*b*) Box-like compartments in cork tissue as observed by Robert Hooke.

CELL THEORY

The cell theory was jointly put forward by **Schleiden** and **Schwann** (1939) in their paper *'Microscope investigations on the similarity of structure and growth of animals and plants'*. Schleiden and Schwann compared their findings and discussed Schwann's hypothesis (cell hypothesis) and formulated 'the cell theory'. The theory states that *'the cells are the units of both structure and function of organisms'*. However, the proposers of cell theory were not the first, to note the cells in organisms. Yet, they are credited with cell theory, because they stated it with such clarity and force that the idea gained general acceptance by biologists of the time. This theory, however, was not able to explain, how new cells are formed. In fact, Schleiden thought that new cells were budded off from the nucleus. **Rudolf Virchow** (1838) observed that new cells develop by division of pre-existing cells—*'Omnis cellulae cellula'* (Theory of cell lineage).

The work of Schleiden and Schwann shows how a hypothesis becomes a theory. A hypothesis (as proposed by Schwann), is a statement that needs confirmation by further observations. Confrmation of a hypothesis by years of verification (by Schleiden in this case) leads to a theory. The theory is modified as and when new discoveries are made (as by Virchow in this case).

Fundamental Features of Cells Theory

The cell theory states that :

1. All living things are composed of cells and their products.
2. All cell arise from pre-existing cells.
3. All cells are basically alike in chemical composition and metabolic activities.
4. The cells are structural and functional unit of all living orgrnisms.
5. The functions of an organism as a whole is the outcome of the activities and interaction of the constituent cells.

Cell as a Basic Unit of Life

All life begins as a single cell. The unicellular organisms complete their entire life cycle as a single cell whereas the cells of multicellular organisms are formed from a single cell-zygote by division.

In a multicellular organism, the cells are not similar but are structurally specialised for specific functions. The cells are grouped into tissues, tissues into organs and organs into organ systems. A group of cells with similar structure and performing similar functions constitute a **tissue.** Several types of tissues collectively form an **organ** which carries out one or more specific activities. In a highly evolved multicellular organism such as man, several interrelated organs constitute an organ system to carry out a particular life process. For example, the organs such as stomach, intestine, liver, pancreas etc. form the digestive system. It is specialised for digestion of food, absorption of digested food and elimination of undigested matter. Similarly, shoot system of a plant is made of stem, branches, leaves, flowers and fruits. It take part in conduction, photosynthesis and reproduction. Organ system is the highest unit of structure and function in an organism. There are several organ systems, each concerned with a separate life function, *e.g.*, digestive system, respiratory system, excretory system, circulatory system, nervous system, reproductive system etc. Several types of organ systems in the body of an organism show unique example of **division of labour**. The organ system together form an organism.

The body of an organism may consists of a number of organ systems, each organ system comprises several organs, each organ consists of many tissues, and each tissue is composed of numerous cells similar in origin and function. A cell is the smallest and basic unit of structure and function in the living beings.

The cells are not only building blocks of the body, they are also the functional unit of life. In the body of an organism a number of different types of cells may coexist. The activities of an organism are sum total of coordinated activities of its constituent cells. All cells in the body have the same genetic material, but the mature cells may become specialised to perform a specific function. All the new cells of an organism develop from the pre-existing ones. Thus, each cell has the same genetic information. It is, therefore, capable of giving rise to a new individual. The ability of a living cell to form a complete organism is called **totipotency.** Each cell is made up of several organelles present within the cytoplasm, just like the ones carried by different organ systems of the body. Thus, all the life activities of an organism are present in miniature form in each cell and every cell of its body. Therefore, the cell can be called a **basic unit of life** and **'the structural unit of an organism'.**

Size, Shape and Lifespan of Cells

The cells exhibit an endless variation in size, shape and lifespan.

Size. Most of the cells are microscopic in size within limits of 3–30 μm. However, some cells are visible to the naked eyes, *e.g.,* an ostrich egg. It is the largest animal cell which measures 170 × 135 mm. In general, eggs are large sized cells because they store food for partial or complete development of embryo. The nerve cell of human being has been found to have the length of 90–100 cm. The smallest known cell is that of *Mycoplasma laidlaivia* —a mycoplasm or PPLO (pleuropneumonia like organism), which measures 0.1–0.5 μm (1000–5000 times the diameter of hydrogen atom). The single called alga, *Acetabularia* measures nearly 10 cm in height. The cells of some plant fibres (*e.g.*, manila hemp) may attain a length of over 100 cm.

The cells generally remains small in size so that these have a large surface area, but small volume to maintain a large surface area-to-volume ratio, which is required for the efficiency of cellular activities such as absorption of nutrients exchange of gases with the environment and metabolism.

Shape. Cells also vary greatly in their shape. These may be disc like, polygonal, rectangular, cuboid, thread like, branchial and even irregular in shape (Fig. 2.2). The shape of the cell may also vary with the function they have to perform. For example,

— the human RBCs are circular and biconcave to transport oxygen.

— the WBCs are amoeboid to squeeze out through the capillary walls.

— the nerve cells are elongated to conduct impulses from different parts of the body to the brain.

— the columnar epithelial cells which form the lining of the intestine are tall and narrow, their height exceed the width.

— the palisade cells in plants are rectangular.

— the guard cells of the stomata are bean shaped.

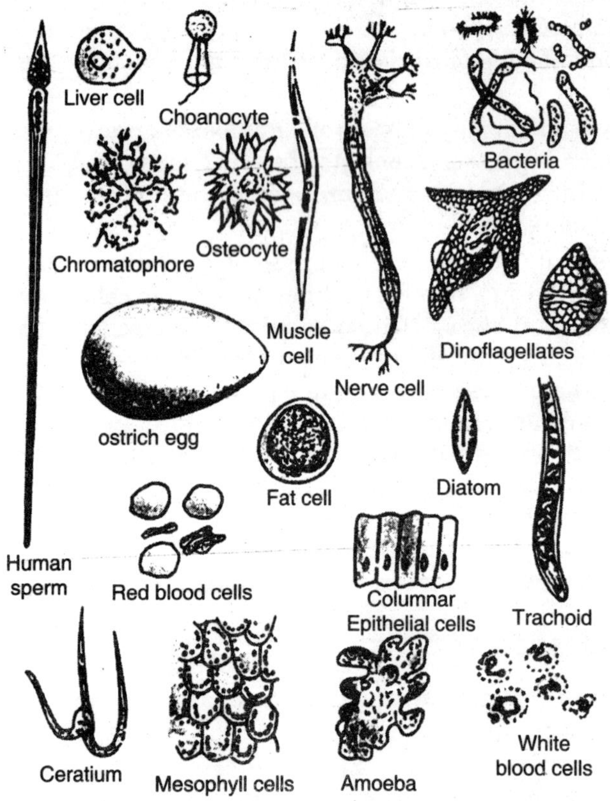

Fig. 2.2. Diagram showing different shapes of the cells.

STRUCTURE OF CELL

On the basis of organisation of genetic material, the cells are of two types : prokaryotic cells and eukaryotic cells. (Fig. 2.3 and Fig. 2.4)

1. **Prokaryotic cells** (Gk. *Pro* = primitive, *karyon* = nucleus). The cells of bacteria and cyanobacteria (blue green algae) lack a well organised nucleus and are called **prokaryotic cells.** Their DNA is not enclosed by a nuclear membrane. They also lack membrane bound organelles. The organisms possessing prokaryotic cells are called **prokaryotes** or **prokaryotic organisms.** Prokaryotes are considered to be primitive organisms.

2. **Eukaryotic cells** (Gr. *Eu* = good, *karyon* = nucleus). The cells of plants and animals possess a well organised nucleus and are called **eukaryotic cells.** Their genetic material is enclosed within a nuclear membrane. The organisms possessing eukaryotic cells are called **eukaryotes** or **eukaryotic organisms.**

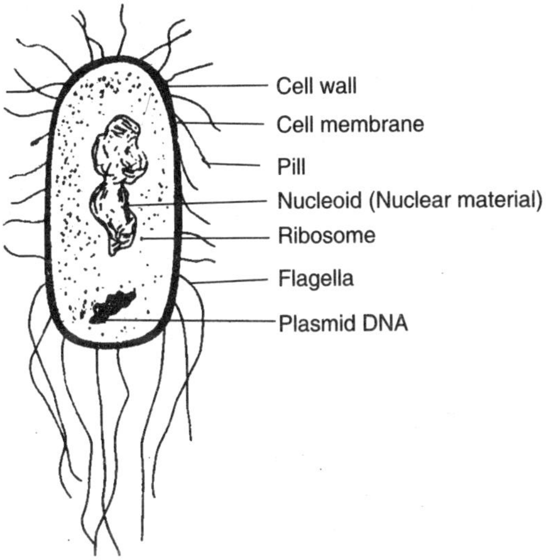

Fig. 2.3. A prokaryotic cell as seen in the electron microscope.

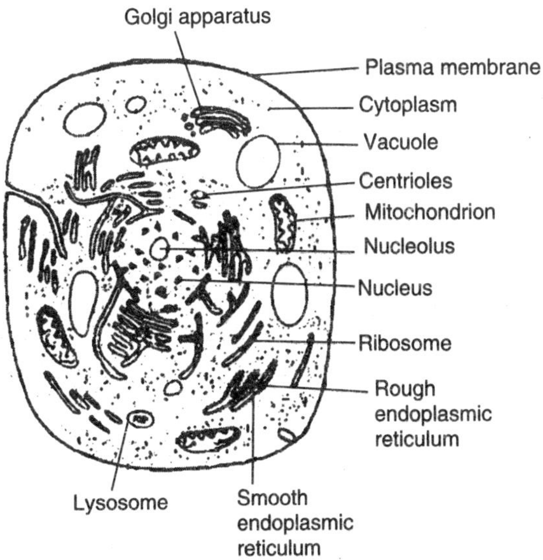

Fig. 2.4. Ultra structure of a typical animal cell (an Eukaryotic Cell) as seen in the electron microscope.

Differences between Prokaryotic Cell and Eukaryotic Cell

Prokaryotic Cell	*Eukaryotic Cell*
1. It is generally smaller (1–10 mm) in size. (1 mm = 10^{-6} m)	1. It is comparatively larger (5–100 mm) in size.
2. It lacks a well organised nucleus as its nuclear material is not surrounded by a nuclear membrane.	2. It contains a well organised nucleus as its nuclear material is surrounded by a nuclear membrane.
3. It has a single chromosome.	3. It contains more than one chromosome.
4. Nucleolus is absent.	4. Nucleolus is present.
5. It lacks membrane bound cell organelles.	5. It possesses membrane bound cell organelles.
6. Cell division occurs by fission or budding. Mitotic and meiotic divisions are absent.	6. Cell division takes place by mitosis and meiosis.

Eukaryotic Cell and its Organisation

The eukaryotic cells have a true nucleus, which contains DNA enclosed by a double membrane envelope. They have a more elaborate internal organisation than the prokaryotic cells. The eukaryotic cells occur in some protists, fungi, plants and animals.

The cells of protists, fungi and plants consist of **cell wall** and **protoplast.** Animal cells lack cell wall. The protoplast denotes the whole of the protoplasm present in the cell. It is differentiated into **plasma membrane** (cell membrane or plasmalemma), **cytoplasm, nucleus** and **vacuoles.** The cytoplasm consists of a jelly like colloidal, ground substance called **cytoplasmic matrix** or **hyaloplasm.** A variety of structures called **organelles** (organoids) and **inclusions** are suspended in it. The cytoplasmic matrix alongwith the structures it contains, is often in constant motion. This phenomenon is called **streaming movement** or **cyclosis.**

Cell organelles are organised protoplasmic subunits having specific functions. They are endoplasmic reticulum, plastids, mitochondria, golgi bodies, centrioles (central apparatus or centrosome), ribosomes, lysosomes, sphaerosomes, peroxysomes, glyoxysomes, microtubules, microfilaments, etc.

Cell inclusions include starch grains, glycogen granules, fat droplets, aleurone grains, excretory or secretory products and crystals.

Nucleus is a large spherical body embedded in the cytoplasmic matrix. It is bounded by a double membrane envelope and contains nucleoplasm, chromatin and one or more nucleoli.

Vacuoles are fluid filled membraneous vesicles occuring in the cytoplasm. In mature plant cells, a large central vacuole is present in the centre of the cell, while the protoplasm occur as a thin layer just inside the plasma membrane. The membrane around the vacuole is called **tonoplast.**

The main functions of different parts of the eukaryotic cell are given below :

1. **Cell Wall.** Provides shape, rigidity and protection to cell.
2. **Cell Membrane.** Regulates entry and exit of substances into and out of the cell.
3. **Cytoplasm.** It contains following organelles.

(*i*) ***Endoplasmic Reticulum.*** Provides cytoskeleton and channelisation, helps in synthesis of fats, steroids and proteins and formation of vacuoles and vesicles.

(*ii*) ***Chloroplasts.*** Photosynthesis.

(*iii*) ***Leucoplasts.*** Storage of food (starch fat or protein).

(*iv*) ***Chromoplasts.*** Provide colour to flowers and fruits.

(*v*) ***Mitochondria.*** Aerobic respiration (Kreb's cycle), synthesis of amino acids and fatty acids.

(*vi*) ***Golgi Apparatus.*** Storage, secretion and excretion, wall synthesis, lysosome formation, membrane transformation and transformation of certain chemicals.

(*vii*) ***Ribosomes.*** Protein synthesis.

(*viii*) ***Centrioles.*** Formation of astral poles and flagella.

(*ix*) ***Lysosomes.*** Separation and storage of hydrolytic (digestive) enzymes, digestion, autophagy.

(*x*) ***Sphaerosomes.*** Metabolism, storage and synthesis of fat.

(*xi*) ***Glyoxysomes.*** Glyoxylate cycle, conversion of fat to carbohydrates.

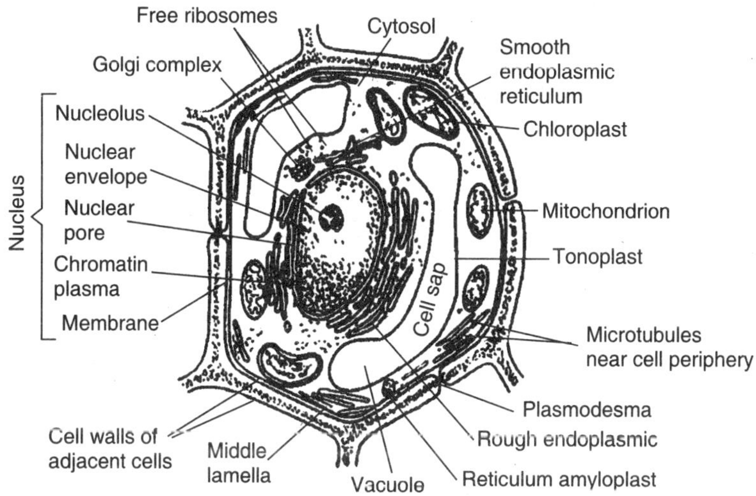

Fig. 2.5. A generalised ultra structure of an eukaryotic plant cell.

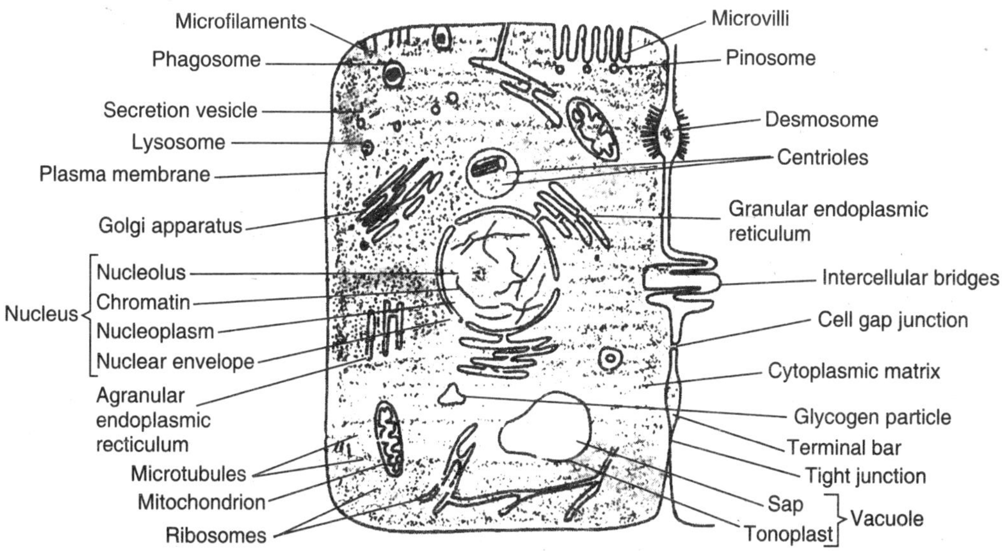

Fig. 2.6. A generalised ultra structure of a eukaryotic cell without cell wall (animal cell).

(*xii*) ***Peroxysomes.*** Photorespiration, peroxide metabolism.

(*xiii*) ***Microtubules.*** Cytoskeleton, formation of spindle and flagella.

(*xiv*) ***Microfilaments.*** Holding of membrane proteins, controlling of cleavage and cyclosis.

(*xv*) ***Vacuoles.*** Maintain osmotic pressure and storage.

4. **Nucleus.** Contains hereditary information, controls cell metabolism, cell differentiation and reproduction, synthesis of DNA and RNA, formation of ribosomes.

Though the cells of plants (Fig. 2.5) and animals (Fig. 2.6) are eukaryotic, even they differ in following respects.

Differences between Plant and Animal Cells

Plant Cell	*Animal Cell*
1. A plant cell is usually larger in size.	1. An animal cell is comparatively smaller in size.
2. It is enclosed by a rigid cell wall in addition to plasma membrane.	2. It is enclosed by a thin, flexible plasma membrane only.
3. It cannot change its shape.	3. An animal cell can often change its shape.
4. Plastids are present. Plant cells exposed to sunlight contain chloroplast.	4. Plastids are usually absent.
5. A mature plant cell contains a large central vacuole.	5. An animal cell often possesses many small vacuoles.
6. Nucleus lies on one side in the peripheral cytoplasm.	6. Nucleus usually lies in the centre.
7. Centrioles are usually absent, except in motile cells of lower plants.	7. Centrioles are practically present in all animal cells.
8. Golgi apparatus consists of a number of distinct or unconnected units called dictyosomes.	8. Golgi aparatus is either localised or consists of a well connected single complex.
9. Lysosomes are rare. Their activity is performed by specialised vacuoles.	9. Lysosomes are always present in animal cells.
10. Glyoxysomes may be present.	10. They are absent.
11. Tight junctions and desmosomes are lacking.	11. Tight junctions and desmosomes are present between cells.
12. Reserve food is generally in the form of starch or fat.	12. Reserve food is usually in the form of glycogen and fat.
13. Plant cell can synthesize all the amino acids, coenzymes and vitamins required by them.	13. Animal cell cannot synthesize all the ammino acids, coenzymes and vitamins required by them.
14. Crystals of inorganic substances may occur inside the cell.	14. Crystals are usually absent.
15. Spindle formed during cell division is anastral *i.e.,* without asters at the poles.	15. Spindle formed during cell division is amphiastral *i.e.*, has an aster at each pole.
16. Cytokinesis occurs by cell plate method.	16. Cytokinesis occurs by constriction or furrowing.
17. Plant cell do not burst if placed in hypotonic solution due to the presence of cell wall.	17. Animal cell lacking contractile vacuoles usually burst, if placed in hypotonic solution.

STRUCTURAL ORGANISATION OF THE CELL

Cell Wall

The cells of some protists, fungi, algae and plants have a protective semi-transparent covering outside the cell called **cell wall.** It is a specialized form of extra cellular matrix which is closely associated with the **external** surface of the cell membrane. The cell wall is sufficiently thick, strong and rigid and measures 0.1 nm to several micrometres in thickness. The cell wall was first seen in the cork cells by **Robert Hooke** in 1665.

Chemical Composition

The composition and appearance of the cell wall differs according to cell type and function it performs.

- The **fungal cell** walls have a thick layer of polysaccharide fibres composed of either chitin or cellulose and a thin outer layer of mixed glycans.
- The substances commonly found in the cell walls of algae are cellulose, galactans, mannans and minerals such as silicon dioxide and calcium carbonate.
- In higher plants, the cell wall fibres are made up of the polysaccharide-cellulose and are embedded in a highly cross linked matrix of polysaccharides such as pectin, lignin and hemicellulose. Certain other compounds like cutin, suberin, xylen silica, minerals, waxes, tannins, resins and gums also occur in the cell wall of higher plants,

Structure of Cell Wall

Cell wall is differentiated into three parts—middle lamella, primary wall and secondary wall.

1. **Middle Lamella.** It is a thin amorphous cementing layer between two adjacent cells. Middle lamella is just like the common brick wall between two rooms. It is absent to the outerside of surface cells is composed of calcium and magnesium pectates. During ripening of the fruits, the pectic compound of the middle lamella solublize to a jelly like material making the fruits soft.

2. **Primary Wall.** It is the first formed wall of the cell which is produced inner to the middle lamella. Primary wall is commonly thin (0.1–3.0 μm in thickness), elastic and extensible in growing cells. It grows by addition of more wall material within the existing one. Such a growth is termed as **intussusception.** Some cells, such as meristematic and parenchymatous cells possess only primary wall.

Primary wall consists of a loose network of cellulose microfibrils embedded in an amorphous gel like matrix or ground substance. In majority of plants, the microfibrils are made of cellulose. They are made of a polymer of acetyl glucosamine or fungus cellulose in fungi.

The matrix of the wall is chiefly composed of water, hemicellulose, pectin and glycoproteins. Pectin is the filler substance of the matrix. Hemicellulose binds microfibrils with the matrix. While glycoproteins control orientation of microfibrils.

3. **Secondary Wall.** After a cell has reached maturity, a thick secondary wall is laid inner to the primary wall by accretion or deposition of materials over the surface of existing structure.

It is laid down in succession of at least three layers, which are often named S_1, S_2 and S_3 (Fig. 2.7).

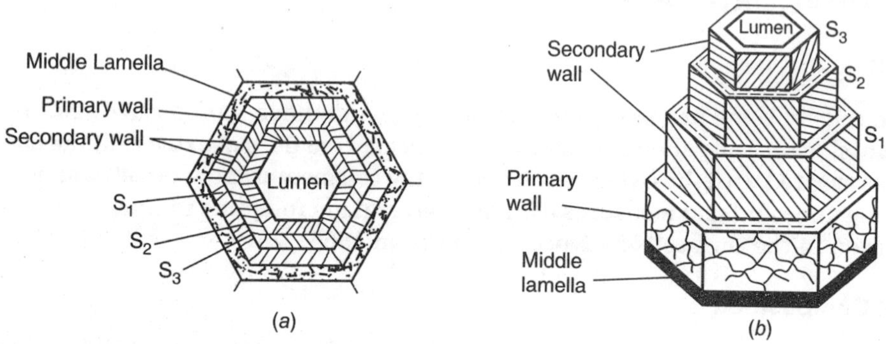

Fig. 2.7. Parts and layers of cells wall; (*a*) a cell in T.S. showing parts of cell wall and (*b*) typical wood fibre cut at various levels to show parts and layers of the wall.

The formation of secondary wall is not uniform in all the cells. This results in the differentiation of various types of cells, such as parenchyma, collenchyma, scierenchyma, fibres and tracheids.

The cellulose microfibrils of secondary wall are compactly arranged with different orientation in different layers in a matrix of pectin and hemicellulose . A number of different materials may be deposited in the secondary wall. **Lignin** is deposited in the secondary wall of xylem and sclerenchymatous cells. The strengthening of cell wall by the deposition of lignin is called **lignification.** The wall of cork and endodermal cells is impregnated with a fatty substance **suberin.** It makes the wall impermeable to water. Other substance which may be deposited in the cells walls are minerals, waxes, tannins, resins, gums etc.

Plasmodesmata. The cell wall does not totally isolate a plant cell. It is interrupted by narrow pores carrying fine strands of cytoplasm, which interlink the contents of the cells. They are callod **plasmodesma connections** or **Plasmodesmata.** They form a protoplasmic continum called **symplast.** A plasmodesma consists of a canal, lined by plasma membrane and having a simple or branched tubule known as desmotubule. Desmotubule is an extension of endoplasmic reticulum. Various substances can pass from one cell to another through plasmodesmata.

Functions of Cell Wall

The cell wall performs a number of functions :

- It provides definite shape to the cell and protects the protoplasm against mechanical injury.
- It gives rigidity and support to the tissues of the plant body.
- It gives strength to the land plants to withstand gravitational forces.
- It counteracts osmotic pressure and prevents bursting of plant cells due to endosmosis.
- By its growth, the wall help in coll expansion.
- It has some enzymatic activity connected with metabolism.
- In many cases, wall takes part in offence and defence.
- Pits present in the wall permit a protoplasmic continuum or symplast amongst cells.
- Cutin and suberin of the cell wall reduce the loss of water through transpiration.
- The walls of xylem vessels, tracheids and sieve tubes are specialised for long distance transport.

Cell Membrane

It is the outer membranous covering of the protoplast or cell protoplasm. Cell membrane is also called **plasma membrane** or **plasmalemma.** It forms a barrier between the external environment and internal organisation of the cell. The presence of such a limiting boundary is the essential feature of all the living cells whether it is prokaryote or the eukaryote. This limiting boundary of the cell is so thin that it could not be resolved and studied upto late 1950s.

Plasma membrane is flexible and is made up of a continuous bilayer of lipid molecules with protein molecules embedded in it.

Fluid Mosaic Model of Plasma Membrane. Based on chemical analysis and physical properties **S.J. Singer** and **G. Nicholson** (1972) proposed a widely accepted **"fluid mosaic model"** of cell membrane. According to this model, the plasma membrane or any membrane of the cell is of quasi fluid nature and is made up of a phospholipid bilayer and two type of protein molecules floating about in the fluid phospholipid bilayer (Fig. 2.8).

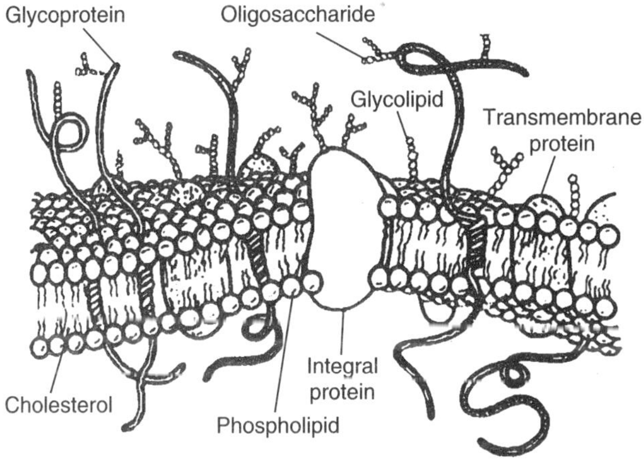

Fig. 2.8. Fluid mosaic model of a membrane having glycocalyx or carbohydrates attached to outer proteins and lipids.

Depending on the degree of association and the ability of separation, the membrane proteins can be classified into two types : peripheral (extrinsic) and integral (intrinsic).

The **peripheral** or **extrinsic** proteins can be dissociated easily with a mild treatment with salt concentration. These proteins are present entirely outside the lipid bilayer. The peripheral proteins form weak electrostatic bonds with the polar functional group of the membrane lipid or the exposed hydrophilic domains of the transmembrane integral protein.

The **integral** or **intrinsic** proteins account for 70% of the total membrane proteins. They cannot be extracted from the membrane without disrupting the latter. They can only be separated by drastic method of extraction like treatment with soap solution or detergents. The distribution of the peripheral as well as integral proteins is highly asymmetrical. The integral proteins pass into the lipid bilayer to different depths. Some of them run throughout the lipid bilayer. They are called **tunnel proteins** or **transmembrane proteins.** Transmembrane proteins may extend beyond the two surfaces as a single helix (*e.g.*, glycophorins). The tunnel proteins individually or in a group form channels for the passage of water and water-soluble substances. The channels, however, possess selective properties for passage of different ions and other polar substances. The proteins are held in their position by both polar (hydrophilic heads of lipids) and non-polar (hydrophobic (tails of lipid) side chains.

Many membrane proteins function as enzymes. Some of them behave as **permeases** for allowing facilitated diffusion. A few proteins act as **carriers** because they actively transport different substances across the membrane. Depending upon their role in active transport carrier can be **uniporters, symporters** and **antiporters.** Certain other proteins function as **receptors** for hormones, recognition centre and antigens. Some lipids and extrinsic proteins present on the either side possess small carbohydrate molecules to form **glycolipids** and **glycoproteins.** They constitute **glycocalyx** or **cell coat.** Conjugated oligosaccharides function as recognition centres, sites of attachment, antigens, etc.

Functions of Cell Membrane

The cell membrane performs a number of functions :

- The major function of cellular membranes is compartmentalisation. As plasma membranes, they separate the cells from their external environment. As coverings of organelles, they help the latter in maintaining specific internal environment and functional individuality.
- Cell membrane protects the cell from injury and keeps the semifluid cell content in place.
- The membranes allow the flow of materials and information between different organelles of the same cells as well as between one cell and another.
- Plasma membranes of the adjacent cells from various types of junctions to keep the cells together.
- By forming plasmodesmata, the cell membranes establish cytoplasmic continuity between adjacent cells.
- Differential permeability and retentivity of plasma membrane as well as other biomembranes control cell metabolism.
- As microvilli, the membrane becomes specilised for absorption of various substances.
- The membrane inflods help in the intake of materials by endocytosis.
- Secretory, excretory and waste products are thrown out by plasma membrane through exocytosis.
- Cell membranes contain enzymes for performing certain reaction on their surfaces, *e.g.*, ATPase, esterases, etc.
- Oligosaccharides of cell membranes help the cells in recognising self and non-self.
- Cell membranes has receptors for informational molecules reaching the cell surface.
- Membranes have carrier proteins for active transport.
- Plasmalemma provides sheaths for cilia and flagella.
- In nerve cells the cell membrane takes part in transmission of impulses.
- Plasma membrane helps certain cells in movement by forming pseudopodia (*e.g.*, *Amoeba*) or undulations (*e.g.*, fibroblasts).

MEMBRANE TRANSPORT

The passage of substances across the cell membrane occurs by three methods passive transport, active transport and bulk transport.

Passsive Transport

It is a mode of membrane transport which occurs without the expenditure of cell energy. Passive transport occurs by passive diffusion and facilitated diffusion.

(*i*) ***Passive Diffusion.*** In this type of transport, the cell membrane plays a passive role. It can occur either through lipid matrix of the membrane or with the help of channels (Fig. 2.9).

- ***Neutral solutes*** or ***Lipid soluble substances*** can move across the plasma membrane through simple diffusion along their concentration gradient *i.e.,* from higher concentration to the lower concentration.

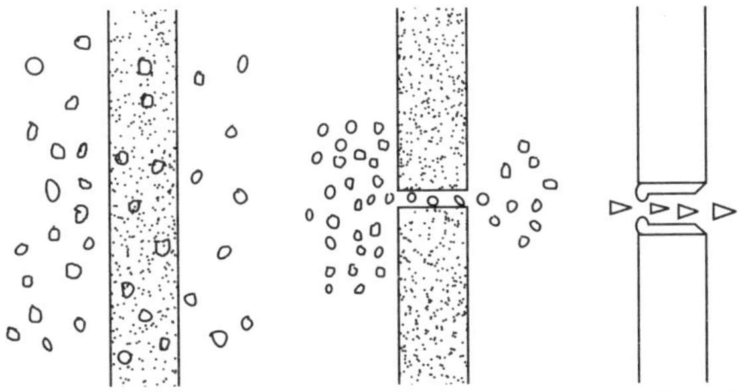

Fig. 2.9. Modes of passive transport.

- ***Channel transport.*** Membranes possess channels in the form of tunnel proteins. There are separate channels for water and different ions. Water channels called, aquaporins allow water and water soluble gases (CO_2 and O_2) to pass through according to their concentration gradient.

Osmosis is an example of such a transport. Diffusion under pressure across a membrane having minute pore is called **filtration.**

Ultrafiltration occurs during glomerular filtration inside kidneys. The process of separation of small particles (*e.g.*, crystalline solutes from larger ones *e.g.*, colloids) due to difference in the rate of diffusion across a membrane having very minute pores is called **dialysis,** *e.g.*, separation of waste products from blood in artificial kidney.

(*ii*) ***Facilitated Diffusion.*** This type of transport occurs through the agency of special membrane proteins called **permeases,** without the expenditure of energy. The transport occurs along the concentration gradient but permeases function as facilitated pathways for the movement of substances. As a result the rate of transport is stereospecific. Transport of glucose into red blood corpuscles is a facilitated diffusion.

Active Transport

It is the movement of materials across the membrane against their concentration or electrochemical gradient. This often leads to accumulation of substances at a higher concentration within the cell than outside. Energy is required for such a process, which is usally provided by ATP. Active transport occurs in case of both ions and non-electrolytes *e.g.*, salt uptake by plant cells, glucose in case of renal tubules, sodium and potassium in case of nerve cells etc.

Active transport is far more rapid than diffusion and is invariably in one direction. There are certain proteins in the membrane, which act as **carrier molecules** or **carrier proteins.** There is a special carrier molecule for each substrate (ion or molecule). It has a binding site for the substrate. The carrier present its binding site on one surface of the membrane. On binding with the substrate the carrier undergoes a conformational change (Fig. 2.10). This transports the carrier bound substrate to the other side of the membrane, where it is discharged in an unaltered state. The carrier is left behind within the membrane. This process needs energy, which is supplied by ATP. In the process ATP is converted to ADP.

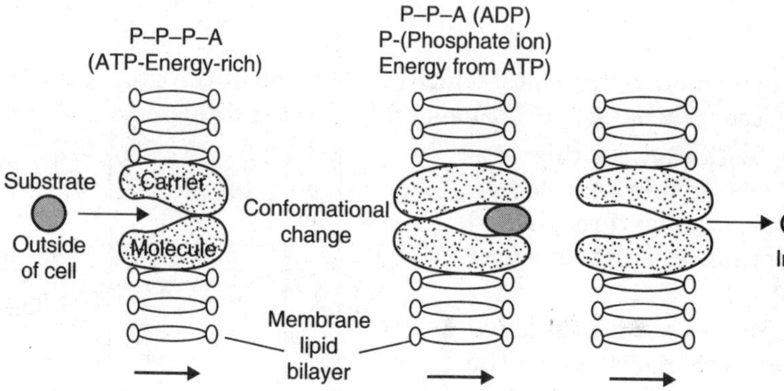

Fig. 2.10. Diagrammatic representation of role of carrier in active membrane transport.

Bulk Transport (Endocytosis and Exocytosis)

The flexibility of the cell membrane enables the cell to engulf food and other material from the external environment and in bulk expulsion of certain material out of the cell. The bulk intake of materials by the cell through cell membrane is called **endocytosis** (Fig. 2.11). Endocytosis is of two types. (*i*) **Pinocytosis** (cell drinking or intake of droplets of extracellular

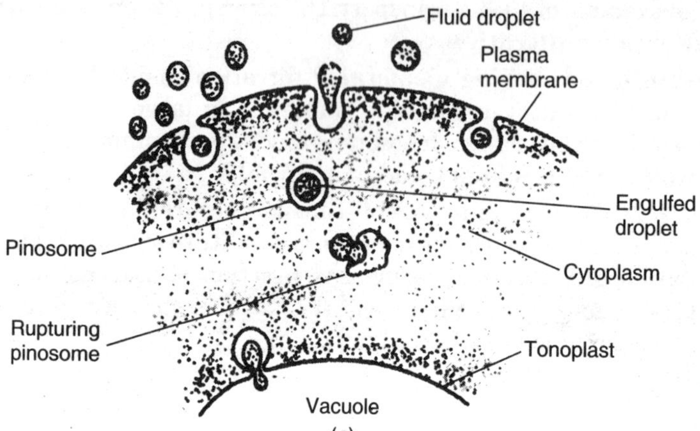

(*a*)

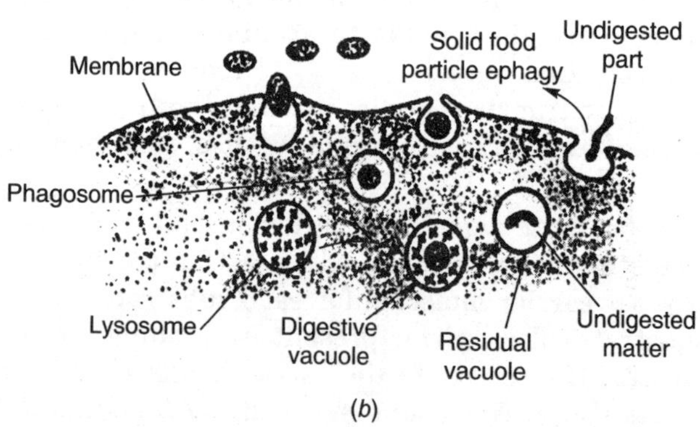

(*b*)

Fig. 2.11. Bulk transport; (*a*) Pinocytosis and (*b*) Phagocytosis.

fluid alongwith sub-microscopic particles forming detachable vesicles called pinosomes) and (*ii*) **Phagocytosis** (cell eating or intake of solid particles such as food, foreign particles, pathogens, etc. forming detachable vesicles called phagosomes). Phagocytosis is frequently used by protozoans (*e.g.*, *Amoeba*) for engulfing food particles, and by specialised mammalian cells called phagocytes (*e.g.*, macrophages and WBC) for removing damaged cells, bacteria and dust particles.

The bulk expulsion of materials from the cell through cell membrane is called **exocytosis.** Exocytosis is quite common in secretory and excretory cells.

Cytoplasm

The term cytoplasm is used for the jelly like fluid general mass of protoplasm excluding the nucleus but including all other components, cytoplasmic matrix, cell organelles and cell inclusions.

A. Cytoplasmic Matrix or Cytosol (Hyaloplasm)

Cytoplasmic matrix is a transparent, homogenous fluid part of the cytoplasm. It can exist in **plasma gel** (viscous) and **plasmasol** (non-viscous) states. Plasma gel is usually present below the plasma membrane and is called **ectoplast.** Plasmasol is internal and is known as **endoplast.** Cytoplasmic martrix is a crystallocolloidal complex in water, where a number of chemicals are present in the form of true solution, while the others are present an colloidal solution. The chemicals present in cytoplasmic matrix include minerals, sugars, amino acids, tRNA, nucleotides, vitamins, proteins, enzymes etc. Proteins occur in colloidal sols.

B. Cell Organelles

They are sub-cellular structures with characteristic morphological forms and definite functions. A cell contains a number of organelles like mitochondria, plastids, endoplasmic reticulum, Golgi complex, lysosomes, microbodies, ribosomes, cytoskeletal structures, cilia and flagella, centrioles etc.

Mitochondria

Mitochondria are membrane bound cell organelles, essential for aerobic respiration of eukaryotic cells. They are popularly called **'power houses'** of cell because they are the major centre cf release of energy. Mitochondria were first observed by **Kolliker** (1880). They were named mitochondria (Gk. *mitos*-thread, *chondrion*-granular) by **Benda** (1897).

The number of mitochondria per cell is variable depending on the physiological activity of the cells. Considerable degree of variability is observed in the shape and size of mitochondria. Typically it is sausage shaped or cylindrical having a diameter of 0.2–1.0 μm (average 0.5 μm) and length 1.0–4.1 μm.

Each mitochondrion is a double membrane bound structure with the outer membrane and the inner membrane dividing its lumen distinctly into two aqueous compartments. *i.e.*, the outer compartment and the inner compartment. The inner compartment is called the **matrix** (Fig. 2.12). The outer membrane forms the continuous limiting boundary of the organelle. The inner membrane forms a number of infoldings called the **cristae** (sig-crista) towards the matrix. The two membranes have their own specific enzymes associated with the mitochondrial function. The matrix also possesses single circular DNA molecule, a few RNA molecules, ribosomes (70s type) and the components required for the synthesis of proteins. The mitochondria divide by fission.

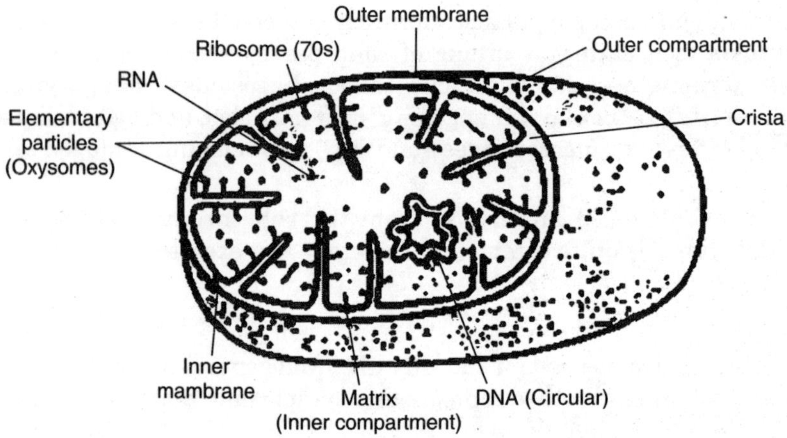

Fig. 2.12. Structure of a mitochondrion (partly cut open to show internal and external structure).

The mitochondria synthesise all the enzymes required for their functioning. Their functioning is partially controlled by the nucleus of the cell and availability of materials from the cytoplasm. Therefore, mitochondria are considered as **semiautonomous organelles.**

Functions of Mitochondria

- Mitochondria are the main seat of aerobic respiration, where respiratory substrates are completely oxidised to carbon dioxide and water. The energy liberated in the process is utilised in the synthesis of energy rich molecules ATP. The latter are used to bring about energy requiring activities of the cell. On this account, mitochondria are called **'power houses'** of the cells.
- Mitochondria provide intermediates for the synthesis of important biomolecules like chlorophyll, cytochromes, steroids etc.
- The matrix of mitochondria contains enzymes for the synthesis of fatty acids.
- Synthesis of many fatty acids occurs in mitochondria.
- Mitochondria regulate calcium ion concentration in the cell by storing and releasing Ca^{2+} as required.

Plastids

Plastids (**E. Haeckel**, 1866) are found in plant cells and some protists. These are large organelles which can be observed under the microscope. Plastids develop from the colourless precursors called protoplastids. They contain specific pigment, thus impart specific colour to certain plant parts.

Based on colour, plastids are of three types–leucoplasts, chromoplasts and chloroplasts.

1. **Leucoplasts** (Gk. *leuko*–white; *plastos*-moulded). They are the colourless plastids of varied shapes and sizes. They possess internal lamelle but lack pigments. Leucoplasts occur in large numbers in the cells of fruits, seeds, tubers and rhizomes, and store food. There are three types of leucoplasts (*i*) **Amyloplasts** store carbohydrates (starch) *e.g.*, potato tubers, wheat and rice (*ii*) **Elaioplasts (Lepidoplasts)** store oil and fats *e.g.*, tuberose and (*iii*) **Aleuroplasts (Poteinoplasts)** store proteins *e.g.*, aleurone cells of maize grain.

2. **Chromoplasts** (Gk. *chroma*-colour; *plastos*-moulded). They are coloured plastids containing pigments other than chlorophyll. Chromoplasts contain fat soluble carotenoid pigments like carotenes, xanthophylls and others. They give the part of the plants (such as flowers, fruits, etc.) a yellow, orange or red colour. They are formed either from leucoplasts or

chloroplasts. Change of colour from green to red during ripening of many fruits (*e.g.*, tomatoes, chillies) is due to transformation of chloroplasts to chromoplasts. Orange colour of carrot is due to the presence of chromoplasts. The latter also provide colour to many flowers and fruits.

3. **Chloroplasts** (Gk. *Chloros*-green, *plastos*-moulded). They are green coloured plastids which possess photosynthetic pigment-chlorophylls and carotenoids responsible for trapping light energy essential of photosynthesis.

Majority of chloroplasts of the green plants are found in the mesophyll cells of the leaves. These are lens shaped, oval, spherical, discoid or even ribbon like organelles having variable length (5–10 mm) and width (2–4 mm). Their number varies 1 per cell of the *Chlamydomonas* (a unicellular green alga) to 20–40 per cell in the mesophyll tissue of leaves.

Like mitochondria, the chloroplasts are also bounded a double membrane envelope. The space limited by the inner membrane of the chloroplast is called the **matrix** or **stroma.** A number of organised flattened membranous sac like structures called **thylakoids**, are present in the stroma (Fig. 2.13). Thylakoids are arranged in stacks like the piles of coins called **grana.** (sig.; granum) or **the intragranal thylakoids.** In addition, there are flat membranous tubules called the **stroma lamellae** (intergranal thylakoids) connecting the thylakoids of the different grana. The membrane of the thylakoids contain photosynthetic pigments. The photosynthetic pigments include chlorophyll *a*, chlorophyll *b*, carotenes and xanthophylls. The stroma of the chloroplast contains enzymes required for the synthesis of carbohydrates and proteins. It also contains small, double stranded circular DNA molecules and ribosomes. The ribosomes of the chloroplasts are smaller (70s type) that the cytoplasmic ribosomes (80s types). New chloroplasts arise either from division of pre-existing ones of the division of their precursors-proplastids.

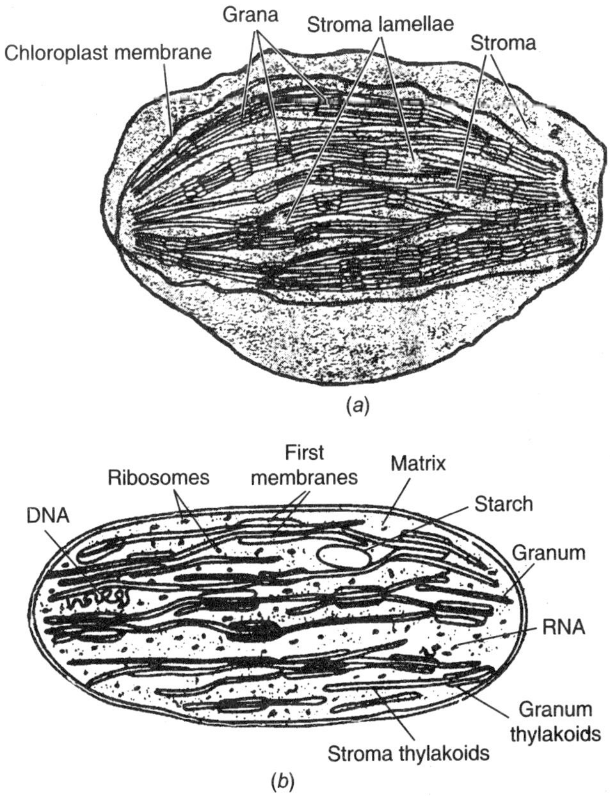

Fig. 2.13. Chloroplast; (*a*) ultra structure and (*b*) Internal structure.

Like mitochondria, chloroplasts synthesize their own enzymes and some of their proteins. However, their functioning in under the control of nucleus of the cell. Therefore, chloroplasts are considered as **'semi-autonomous organelles'.**

Functions of Chloroplast

- The most important and fundamental function of chloroplast is photosynthesis or formation of organic compounds from inorganic raw materials.
- Chloroplasts are able to trap solar energy and change it into chemical energy. This chemical energy is used by all living organisms to perform their life activities.
- Chloroplasts fix up carbon dioxide and release oxygen during photosynthesis. This keeps the percentage of two gases balanced in the atmosphere.
- They store fat in the form of plastoglobuli.
- Chloroplast may synthesise some amino acids and fatty acids.
- They store starch either temporarily (*e.g.*, in higher plants) or permanently (*e.g.*, in several algae).
- Chloroplasts may change into chromoplasts to provide colours to many flowers and fruits to attract animals.

Similarities between Mitochondria and Chloroplast

1. Presence of double membrane envelope, with involutions from inner membranes.
2. Both possess their own DNA, RNA and ribosomes to have functional independence from the cellular machinery.
3. Both are semi-autonomous organelles.
4. They are formed by the division of pre-existing organelles.
5. Both produce ATP.
6. The organelles occur in eukaryotes and are absent in prokaryotes.

Differences between Mitochondria and Chloroplast

Mitochondria	*Chloroplasts*
1. Mitochondria are colourless organelles.	1. Chloroplasts are green organelles.
2. They are found in all types of cells of aerobic organisms, both plants and animals.	2. They are restricted to the green cells of the plants.
3. They are generally cylindrical in outline.	3. They are generally dis-shaped.
4. Their inner membrane is infolded into incomplete partitions called cristae.	4. Their inner membrane gives rise to flattened sacs called thylakoids.
5. The cristae do not produce grana.	5. At places thylakoids produce stacks of discs called grana.
6. Mitochondria do not contain pigments.	6. Photosynthetic pigments-chlorophylls and carotenoids are present in the membrane of thylakoids.
7. The inner membrane and cristae bear large number of elementary particles for ATP synthesis.	7. ATP synthesis is carried out by coupling factors present only on the thylakoids.
8. Mitochondria liberate energy by breaking down of organic food.	8. Chloroplasts store energy by building up organic food.
9. They do not take part in the conversion of light energy into chemical energy.	9. They are the centre of conversion of solar energy into chemical energy.
10. Mitochondria consume oxygen.	10. Chloroplasts liberate oxygen.

Endoplasmic Reticulum (ER)

Electron microscopic studies of eukaryotic cells reveal the presence of a network or reticulum of tiny tubular structures scattered in the cytoplasm. It is called endoplamic reticulum (ER) (Fig. 2.14). Endoplasmic reticulum was first discovered independently by **Porter** (1945) and Thompson (1945). It was named by **Porter** (1953), when he observed that its channels are often concentrated in the endoplasmic region of the cell. Endoplasmic reticulum remains continuous with the plasma membrane, nuclear envelope and Golgi complex. It may develop from pre-existing ER, plasma lemma or nuclear envelope. ER divides the intracellular space into two distinct compartments, *i.e.* luminal (inside ER) and extra luminal (cytoplasm) compartments.

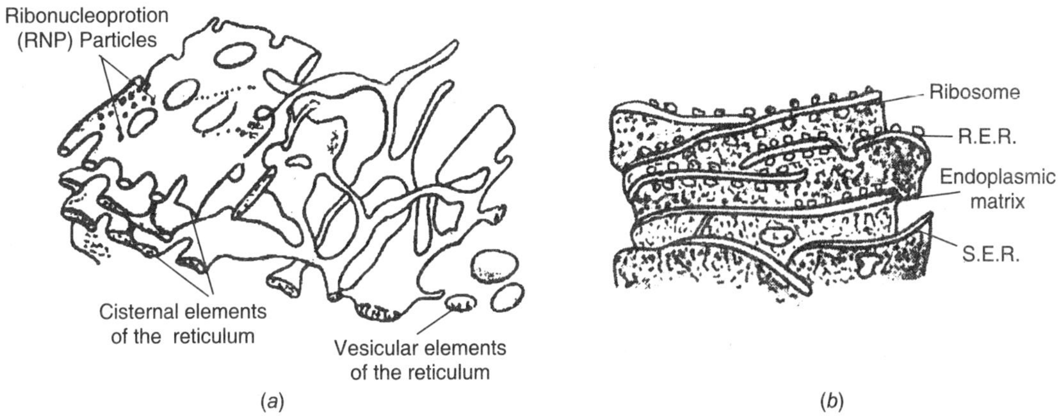

Fig. 2.14. (*a* and *b*) Part of endoplasmic reticulum showing its three dimensional nature.

Depending upon the nature of its membranes, endoplasmic reticulum is of two types-rough (granular) and smooth (agranular).

(*i*) **Rough Endoplasmic Reticulum (RER).** It has membranes with ribosomes attached to the cytoplasmic surface. This makes the surface rough or granular like coarse sand paper.

(*ii*) **Smooth Endoplasmic Reticulum (SER).** It has smooth membranes, which do not bear ribosomes. SER is found in the cells engaged in the synthesis and storage of glycogen, fat and sterols (*e.g.*, liver cells, interstitial cells in testis, adrenal cortex cells, etc).

Functions of ER

RER and SER have some common functions and specialised functions.

Common Functions

- ER provides an ultra structural skeletal framework to the cell and gives support to the colloidal cytoplasmic matrix.
- It provides a large surface inside the cell for various physiological activities.
- ER helps in the rapid intercellular transport of the materials.
- As desmolubulues, it connects two adjacent protoplasts through plasmodesmata.
- As sarcoplasmic reticulum of muscle cells, it conducts impulses from the surface to the deeper parts. In other cells, ER conducts information from cell exterior to inside and form one part of the cell to another.
- ER provides precursors of different secretory substances to golgi apparatus.

- The membranes of ER are the source of nuclear membrane and the membranes of some other organelles like mitochondria and golgi apparatus.
- ER membranes contain a variety of enzymes (*e.g.*, ATPase, dehydrogenases, phosphatases etc.) for various metabolic activities.

Specialised Functions of RER

- RER contains ribophorins for the attachment of ribosomes.
- It offers extensive surface area for synthesis of protein on the ribosomes.
- Proteins and enzymes synthesised by ribosomes enter the channels of RER both for intracellular and extracellular transport.
- It provides enzyme precursors for the formation of lysosomes by golgi apparatus.

Specialised Functions of SER

- SER possesses enzyme bodies called **glycosomes** for glycogen metabolism (glycogenesis and glycogenolysis) in the liver cells.
- It helps in synthesis of fats in the cells of adipose tissue.
- It is involved in the formation of visual pigments from vitamin A in retinal cells.
- SER brings about detoxification in the liver cells *i.e.*, converts harmful materials into harmless ones for excretion by the cell.
- In the interstitial cells of testis and ovary SER helps in the synthesis of steroid hormones.
- SER is involved in the synthesis of ascorbic acid.
- It gives rise to sphaerosomes.

Golgi Complex (Golgi Apparatus)

Golgi complex or Golgi apparatus is a complex protoplasmic structure made up of flat, expanded, intercommunicating cisternae, a network of tubules with vesicles and vacuoles. It is 0.5 μm to 1.0 μm diameter (Fig. 2.15). It was discovered by an Italian scientist **Camillo Golgi** in 1898 in nerve cells of barn owl and cat by means of **metallic impregnation method.** The organelle was named after its discoverer.

Golgi complex occurs in all cells except prokaryotic cells (PPLO, bacteria, cyanobacteria) and some eukaryotic cells such as sieve tubes of plants, spermatocytes of bryophytes and pteridophytes and mammalian red blood corpuscles.

In animal cells, Golgi complex is either single or consists of a single connected complex.

The two conditions are respectively called **localised** (as in most vertebrate cells) and **diffused** (as in most invertebrate cells, liver and nerve cells of vertebrates). In plant cells and fungi, Golgi complex consists of a number of unconnected units called **dictyosomes.** A plant cell generally has 10–20 dictyosomes.

The Golgi cisternae have continuties of luman with each other and with SER tubules. The Golgi cisternae are concentrically arranged near the nucleus with distinct convex *cis* or the forming face and concave *trans* or maturing face. The *cis* and *trans* faces of the organelle are entirely different, but interconnected. The forming face receives secretory materials through transitional vesicles which are pinched off from SER (Fig. 2.16). Small vacuoles containing amorphous granular substance are pinched off from the maturing face of Golgi apparatus. These are called **Golgian vacuoles.** Some of them contain hydrolysing enzymes and act as lysosomes.

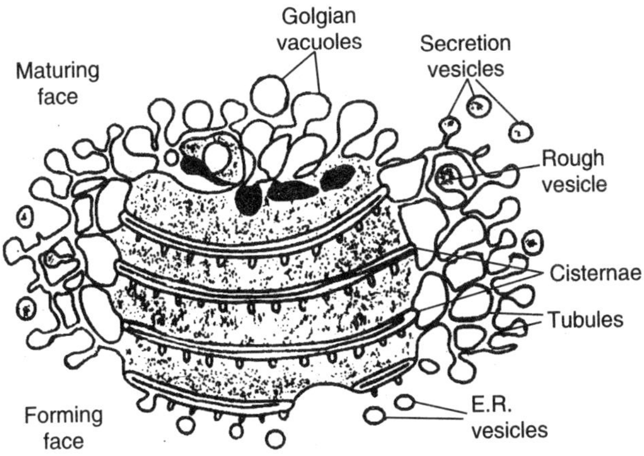

Fig. 2.15. Structure of Golgi apparatus (dictyosome).

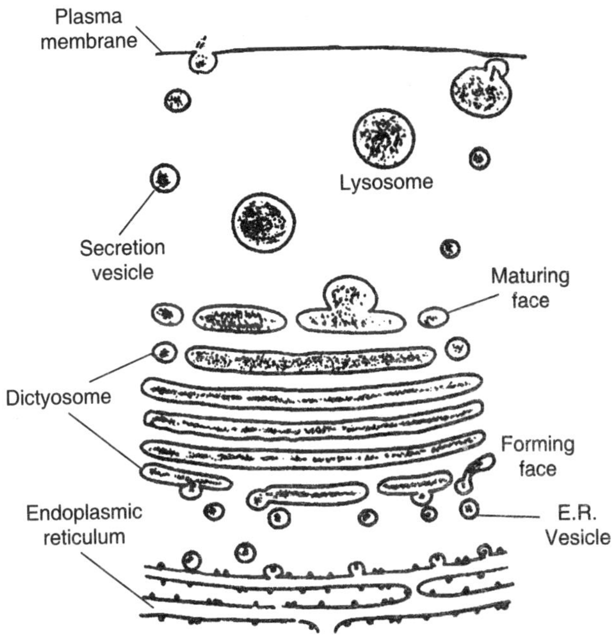

Fig. 2.16. Golgi apparatus or dictyosomal activity in correlation with endoplasmic reticulum.

Functions of Golgi Apparatus

Golgi apparatus perform a number of functions of transformation, synthetic and secretory nature. The important ones are :

- **Secretion.** In glandular cells, Golgi apparatus concentrates their secretions and releases the same outside the cells. The secretory vesicles of the glands usually contains zymogen granules. They are sent out of the cells through exocytosis or reverse pinocytosis.

- **Complex carbohydrates.** Most of the complex carbohydrates, other than starch and glycogen are synthesised from simple sugars inside the Golgi complex. These carbohydrates include pectic compounds, mucopolysaccharides, hyaluronic acid, chondrotin sulphate, hemicellulose, cellulose, etc.
- **Glycoproteins.** Proteins synthesised by the rough endoplasmic reticulum are passed on to the cisternae of Golgi complex. Here, they combine with carbohydrates to form glycoproteins. The process is called **glycosylation.**
- **Lipoproteins.** Lipids and proteins coming from endoplasmic reticulum are complexed into Lipoproteins in the Golgi apparatus. This process is termed as **liposylation.**
- **Transformation of membranes.** Golgi complex is involved in membrane transformation, that is conversion of one type of membrane into another type *e.g.*, the membrane of a secretory vesicle becomes a part of cell membrane after discharging their contents. The complex also take part in recycling of plasma membrane.
- **Formation of cell wall.** In plant cells, dictyosomes synthesise pectic compounds of middle lamella and various polysaccharides of the cell wall. They are brought to the area of new wall synthesis by secretion vesicles.
- **Formation of plasmalemma.** The membranes of vesicles produced by dictyosomes join in the region of cytokinesis to produce new plasmalemma.
- **Formation of lysosomes.** Golgi complex gives rise to primary lysosomes by budding from the maturing face of cisternae.
- **Formation of yolk (Vitellogenesis).** In oocytes of animals, Golgi apparatus functions as the centre around which yolk is deposited.
- **Formation of acrosome.** Acrosome is an important part of the tip of animal sperm, which helps in digesting the covering sheath of ovum during fertilization. It is synthesised by Golgi complex.
- **Hormones.** The production of hormones by endocrine glands is mediated through Golgi apparatus.
- **Matrix.** Golgi complex of the cells of connective tissue help in the formation of matrix of the tissue.

Lysosomes

Lysosomes (Gk. *lysis*-digestive, *soma*-body) are small membrane bound vesicles containing hydrolysing enzymes in the form of crystalline or semi-crystalline granules. Lysosomes were first reported by a Belgian scientist **Christain de Duve** in 1955 in rat liver cells through fractionation technique. The organelles were observed under electron microscope by **Navikoff** (1956). He also coined the term, lysosomes for the organelles.

Lysosomes have been recorded to contain about 40 different hydrolytic (digestive) enzymes. All enzymes do not occur in the same lysosome, but there are different sets of enzymes in different types of lysosomes. The important enzymes are acid phosphatases, sulphatases, proteases, nucleases, lipases and glycosidases. These enzymes usually function in acidic medium of pH less than 7. Therefore, they are commonly called **acid hydrolases.** The covering membrane of lysosomes prevent the enzymes from escaping into the cytoplasm and destroying the latter. If the membrane happens to get broken the various cellular constituents would undergo lysis. The covering membrane becomes fragile in the absence of oxygen, or the presence of excess of vitamins A and E, male and female hormones, bile salts, X-rays and ultra violet rays. The membrane is protected from these agencies by cholesterol, cortisone and chloroquinc.

Lysosomes are present in all animal cells except a few cell types *e.g.*, mammalian red blood corpuscles. Lysosomes are also found in protists, fungi and certain plant cells such as root tip cells of maize, cotton and pea seeds. However, they are less prevalent in plant cells than in animal cells. Prokaryotic cells also lack lysosomes.

Lysosomes are generally rounded but can be irregular (*e.g.*, root tip cells) in outline. The diameter varies from 0.2–0.5 μm. The number of lysosomes in a cell varies with the cell type. They are particularly abundant in certain white blood corpuscles and secretory cells of pancreas, spleen, liver, kidneys. Cells of degenerating or regressing tissue (*e.g.*, tadpole's tail) also contain many lysosomes.

Lysosomes do not normally burst in the cytoplasm. All materials are to be acted upon by lysosome enzyme must enter it as the latter is covered by a membrane. The materials to be digested are usually enclosed inside the vacuoles and the vacuoles fuse with the lysosomes for digestion of materials. Lysosomes take part in intracellular digestion of various types of materials of endogenous or exogenous origin. Extracellular digestion can be performed by them under certain conditions. Lysosomes also help in removing various substances including carcinogens.

Lysosomes arise jointly from golgi complex and smooth endoplasmic reticulum. The precursors of hyrolytic enzymes are synthesised on the rough endoplasmic reticulum. The latter transfers them to the forming face of golgi complex through its vesicles. In golgi complex the precursors are changed to enzymes. The enzymes are then packed in larger vesicles, which are pinched off as lysosomes from the maturing face.

Kinds of Lysosomes

Lysosomes exist in more than one morphological forms. This phenomenon is called **polymorphism.** Depending upon their morphology and function, there are four types of lysosomes—primary, secondary, residual bodies and autophagic vacuoles (Fig. 2.17).

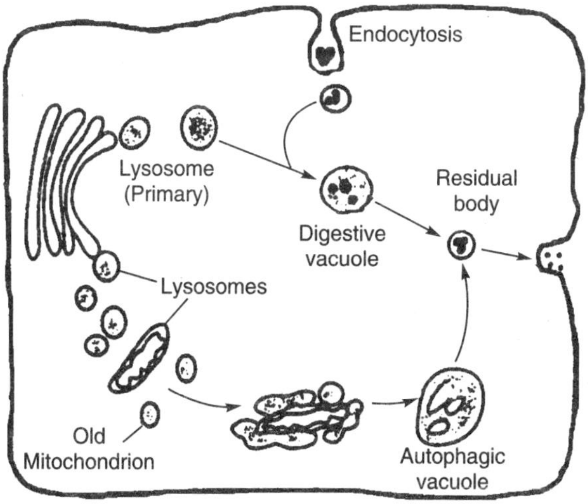

Fig. 2.17. Types of lysosomes and their functions.

(*i*) ***Primary Lysosomes.*** These are newly pinched off vesicles from the Golgi apparatus. They contain hydrolytic enzymes in granular form.

(*ii*) ***Secondary Lysosomes.*** They are formed by the fusion of food containing phagosomes with lysosomes. Secondary lysosomes are also called **heterophagosomes** or **digestive**

vacuoles. In intracellular digestion the primary lysosomes fuse with the vacuoles containing extracellular or intracellular food. The digested food is passes out into the cytoplasm. Finally, the secondary lysosomes is left with undigested food.

(*iii*) ***Residual Bodies (Residual Lysosomes).*** In a secondary lysosome, the enzymes digest the incoming materials. The digested food passes out into the cytoplasm. Ultimately, the secondary lysosome is left with only undigestible matter. The lysosomes which contain only indigestible food materials are called residual bodies. The residual bodies pass outwardly and fuse with the plasma membrane to throw out the debris into external environment by **exocytosis** or **ephagy.** Sometimes residual bodies may remain in the cells due to failure of exocytosis or absence of some hydrolytic enzymes. This contributes to ageing process or certain diseases like hepatitis and polynephritis.

(*iv*) ***Autophagic Vacuoles (Autophagosomes or Autolysomes).*** They are produced by the fusion of a number of primary lysosomes around worn out or degenerate intracellular organelles. The enzymes of lysosomes digest the organelles thus enclosed. This phenomenon is called **autophagy** or **autodigestion.** Besides removing worn out organelles or dead cells, the autophagic vacuoles also help in removing internal obstructions and providing nourishment during starvation.

Functions of Lysosomes

- Lysosomes digest the food contents obtained through phagosomes. It is called **intracellular digestion.**
- They also perform extracellular digestion. For this, lysosomes release their enzymes to the outside through exocytosis.
- Lysosomes of leucocytes take part in **natural defence** of the body by devouring foreign proteins, toxic substances, bacteria and other microorganisms.
- During periods of starvation, lysosomes provide nourishment by rapidly hydrolysing the stored food contents (carbohydrates, lipids and proteins). Extra nourishment may also be obtained by digesting some cellular contents and cells.
- Dead cells and debris that accumulate at the site of injury are destroyed by the lysosomes of some leucocytes. This is called **natural scavenging** of the body.
- Old and useless organelles of the cells are also digested by lysosomes.
- Lysosomes of sperm discharge their enzymes outside the ovum during fertilization for breaking the membrane of the latter.
- Lysosomes are involved in **autolysis** (self-destruction of cell tissue or organ). During autolysis, lysosomes do not enclose the structures to be broken down. Instead, they themselves burst to release the digestive enzymes. Autolysis plays a role in a number of developmental processes such as, the disappearance of tail during the metamorphosis of frog's tadpole, softening of gum tissue to allow eruption of teeth in vertebrates, breakdown of matrix of cartilage for ossification etc. Lysosomes also break down the matrix of the bone during remodelling of the latter in response to injury.
- In thyroid gland, thyroglobulin is hydrolysed by lysosome enzymes to form active thyroxine.
- Lysosomes seem to be essential for cell division perhaps they remove the agents that cause repression of mitotic cycle.
- In certain regions, due to some malfunction, the residual bodies do not undergo exoeytosis. Instead they remain inside the cells and cause disease like hepatitis and polynephritis.

- Lysosomes remove carcinogens by engulfing and separating them.
- The abnormal release of lysosomal enzymes may result in mutations, breakage of chromosomes and other abnormalities. It may lead to blood cancer (leukemia).

> Lysosomes may be called **'suicide bags'** of the cells due to their autolytic role in the cells. They may be called **'disposal unit'** of the cells because they digest the incoming food materials and remove the foreign bodies, toxic substances and debris. They may also be called **'recycling centres'** as they break down worn out cells and cell organelles to component molecules for building new organelles and cells.

Microbodies

They are extremely small membrane bound organelles which utilise molecular oxygen and take part in oxidation other than those involved in aerobic respiration Microbodies are of two types peroxisomes and glyoxysomes.

(*i*) **Peroxisomes.** Peroxisomes are microbodies which contain enzyme for peroxide-biosynthesis. They are found in both plant and animal cells. Peroxisomes are supposed to develop from endoplasmic reticulum. They are generally found in close association in endoplasmic reticulum, mitochondria and chloroplasts. Peroxisomes are spherical or ovoid sacs with 0.5–1 μm in diameter. They are limited by a single membrane and contain a densely granular **matrix.**

Peroxisomes contain peroxide producing enzymes (oxidases) and peroxide destroying enzyme (catalyase). Oxidases catalyse the oxidation of substrates like amino acids and uric acid, producing hydrogen peroxide (H_2O_2). Catalase degrades hydrogen peroxide to water and oxygen. Hydrogen peroxide is harmful to living cells. Its destruction by catalase is, thus, beneficial for the cell.

Plant peroxisomes are found in photosynthetic cells and are associated with chloroplasts. They are involved in **photorespiration.**

In animal cells, peroxisomes help in lipid metabolism and detoxification of alcohol.

(*ii*) **Glyoxysomes.** Glyoxysomes are microbodies which contain enzymes for β-oxidation of fatty acids and glyoxylate cycle. They are quite common in the cells of germinating oil seeds (*e.g.*, castor, watermelon, peanut, etc.) and some fungi. Like other microbodies, glyoxysomes have a covering membrane and a matrix.

Ribosomes (Palade Particles)

Ribosomes are the granular structures first observed under the electron microscope as dense particles by **George Palade** (1953). They are composed of ribonucleic acid (RNA) and proteins, and are not surrounded by any membrane.

Ribosomes occur free in the cytoplasm as well as attached to the our surface of the rough E.R. Each ribosome consists of two subunits—a small and a large subunit (Fig. 2.18). The two subunits join to form complete ribosome only at the time of translation (protein synthesis). At the time of active protein synthesis many ribosomes bind to a messenger RNA

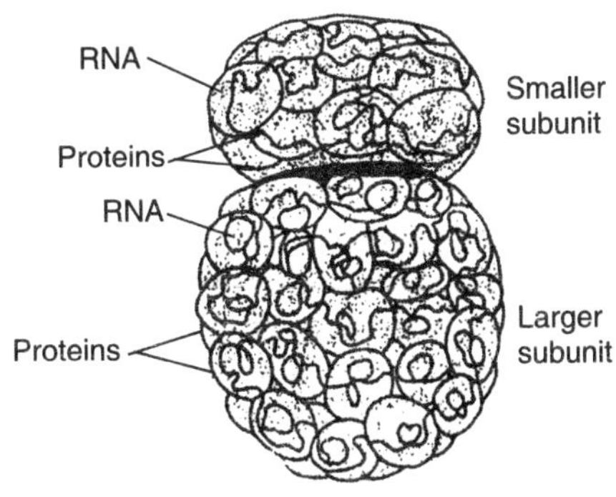

Fig. 2.18. A complete ribosome.

(mRNA) to synthesis many copies of a particular polypeptide (protein). Such a string of ribosomes is called **polyribosome** or **polysome.**

Depending upon size or sedimentation coefficient, ribosomes are of two types: 70s and 80s. Here 's' (Svedberg unit) stands for the sedimentation coefficient. It indirectly is a measure of density and size.

70s ribosomes occur in prokaryotic cells and in semi-autonomous organelles (*i.e.*, mitochondria and plastids) of eukaryotic cells. 80s ribosomes occur in the cytoplasm of eukaryotic cells.

Functions of Ribosomes

Ribosomes are the sites for polypeptide or protein synthesis. Hence, they are known as **protein factories** of the cell.

Cytoskeleton

An elaborate network of filamentous proteinaceous structures present in the cytoplasm is collectively called, the cytoskeleton.

The cytoskeleton in a cell are involved in many functions such as mechanical support, motility and maintenance of cell shape.

Cilia and Flagella

Cilia (sig.: cilium) and flagella (sig.: flagellum) are fine hair like out growths of the cell membrane. Cilia are small structures which work like oars producing a current in the fluid medium for locomotion and passage of substances. Flagella are comparatively longer and responsible for cell movement.

The prokaryotic cells may also possess flagella, but these are structurally different from that of eukaryotic flagella.

The electron microscopic study of a cilium or flagellum show that they are covered with plasma membrane. The central core called **axoneme,** consists of microtubules running parallel to the long axis. The axoneme has a pair of doublets of radially arranged peripheral microtubules and a pair of centrally located microtubles. Such an arrangement of axonemal microtubles is referred to as the 9 + 2 array (Fig. 2.19). The central tubules are connected by

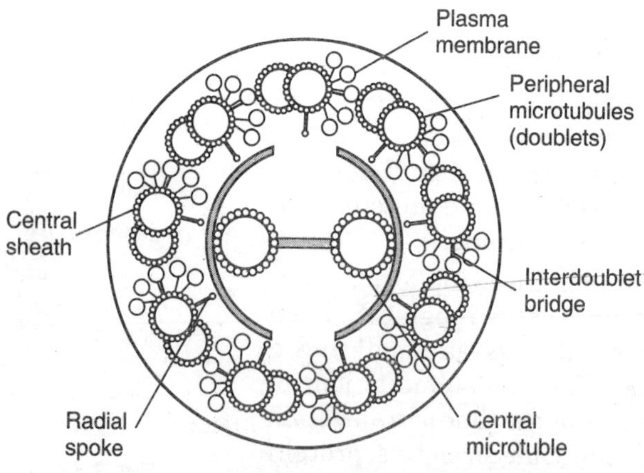

Fig. 2.19. Structure of a cilium/flagellum in T.S.

bridges and is also enclosed by a central sheath. One of the tubules of each peripheral doublets is connected to the central sheath by a radial spoke. Thus, there are nine radial spokes. The peripheral doublets are also interconnected by linkers. Both the cilium and flagellum arise from centriole like structure called the basal bodies.

Centrosome and Centrioles

Centrosome is an organelle found in animal cells. It consists of two cylindrical structures called **centrioles** embedded in a amorphous pericentriolar material. The two centrioles lie perpendicular to each other. The pair of centrioles is often called **diplosome.** The area of cytoplasm in which diplosome lies, called centrosphere. The complex formed of centrioles and centrosphere is called **centrosome** or **central apparatus.**

Each centriole has a cartwheel like organisation. It is made up of nine evenly spaced peripheral fibrils of tubulin protein. Each of the peripheral fibril is a triplet. The adjacent triplets are linked with each other. The central part of the proximal region of the centriole is proteinaceous and forms the hub. The hub is connected with tubules of the peripheral triplets by radial spokes made up of protein (Fig. 2.20).

The centrioles form the basal body of cilia or flagella and spindle fibres forming spindle apparatus during cell division in animal cells.

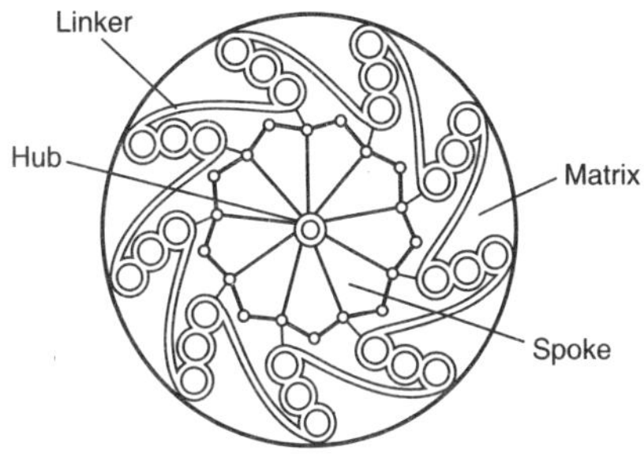

Fig. 2.20. A centriole as seen in T.S.

Vacuoles

Vacuoles are membrane bound sac like vesicles present in the cytoplasm. The vacuolar membrane is similar to plasma membrane and is called tonoplast.

- In animal cells, the vacuoles are smaller in size and fewer in number. They may contain food material and are called **food vacuoles.**
- In unicellullar animals, they take part in osmoregulation and excretion and are called **contractile vacuoles.**
- Plant cells have a large vacuole filled with a fluid called cell sap or vacuolar sap. It is called **sap vacuole.** The cell sap contains mineral salts, sugars, amino acids, proteins, waste products and water soluble pigments.

In a mature plant cell, all the vacuoles fuse to form a single large vacuole called **central vacuole.** It occupies about 90 per cent of the volume of a cell. The cytoplasm of the cell continuously move around the central vacuole. It is called **cyclosis** or **streaming movement.**

- Some prokaryotes have a group of submicroscopic vesicles containing metabolic gases, called **gas vacuoles.** They regulate buoyancy of the cell, provide mechanical support and protect the cell from harmful radiations.

Cell Inclusions

Cell inclusions are non-living substances present in the cells. They are also called **ergastic bodies** or **deutoplastic substances.** These include both organic storage materials and inorganic crystals. The cell inclusions can be grouped into four categories:

(*i*) **Reserve foods** include different types of food reserves such as starch grains, glycogen granules, fat droplets and aleurone (protein) grains.

(*ii*) **Excretory** or **Secretory products** include mucus in several animal cells and essential oils, alkaloids, resins, gums, tannins, latex etc., in plant cells.

(*iii*) **Inorganic crystals** include crystals or incrustations of silica, calcium carbonate, calcium sulphate and calcium oxalate that occur either in cell cavity or in cell wall of plant cells.

(*iv*) **Pigment granules** occur special cells called chromatophores, present in the skin of vertebrate animals.

NUCLEUS

Nucleus (L. *nucleus*-Kernel) *is the most conspicuous and the principal organelle of the cell, which contains all the genetic information for controlling metabolism and transmission to the posterity.* A nucleus in the non-dividing cell or metabolic phase is called **interphase nucleus.** (Fig. 2.21)

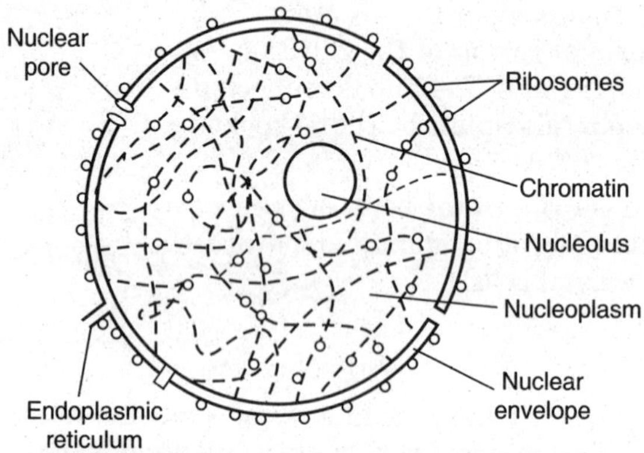

Fig. 2.21. Ultrastructure of interphase nucleus.

Nucleus was first discovered in plant cell by **Robert Brown** (1831). A nucleus is present in all eukaryotic cells except a few cell types such as mature sieve cells of vascular plants and red blood corpuscles of mammals. Even these cells also possess nucleus during the early stages of their development. Prokaryotic cells do not have an organised nucleus. They have a nucleoid (incipient nucleus) represented by a highly folded circular DNA, without a membrane.

Most cells possess single nucleus and are called **uninucleate.** Certain protists such as *Paramecium* are **binucleate** *i.e.,* have two nuclei, one for controlling metabolic activities of the organism and the other possessing hereditary information. The cells of bone marrow, straited muscles, latex vessels, certain algae and fungi possess many nuclei and are called **multinucleate.** The multinucleate animal cells are formed by cell fusion. Such cells are commonly termed as **syncytial cells** Certain fungi *(e.g., Rhizopus)* and algae (*e.g., Vaucheria*) have many nuclei in their large body which is not divided into cells. This condition is called **coenocytic.**

Nucleus has a double membrane envelope—the **nuclear envelope,** which encloses a ground substance called **nucleoplasm** or **karyolymph.** The nuclear envelope possesses many pores called **nuclear pores.** They maintain continuity between the nucleoplasm and cytoplasm. The nucleoplasm has two types of nuclear structures—the **nucleolus** and **chromatin.** The **nucleolus** is a spherical granular body, rich in protein and RNA (ribonucleic acid). It is the site of ribosome formation. Therefore, nucleolus is also called as **factory of ribosomes.** There may be one or more nucleoli in the nucleoplasm.

The **chromatin** is a network of fine threads composed of genetic material. DNA (deoxyribose nucleic acid) and proteins (*i.e.*, histones). During cell division, chromatin is condensed into thick cord like structures called **chromosomes.** The chromosomes contain many segments called **genes** and each gene is responsible for one hereditary character of the organism. They contain information for inheritance of features from parents to next generation in the form of DNA molecule. That is why, the study of heredity or inheritance of traits is called **genetics.**

Functions

- Nucleus contains all the genetic information required for the growth, development, reproduction, metabolism and behaviour of the organism.
- It stores hereditary information, which is passed to the daughter cells by division and to the next generation through the gametes (*i.e.*, sperms and eggs).
- It controls, cell metabolism and other activities of the cell, hence it is called master of the cell.
- All variations that contribute in evolution are caused by the change in the genetic material present in the nucleus.

Structure of Chromosome

Chromosomes are thread like condenced chromatin fibres which contain hereditary information and are visible only during cell division.

The structure of a chromosome can be studied at metaphase stage of cell division. Each chromosome at metaphase stage possesses two halves called **chromatids**, (Fig. 2.22) which lie side by side along their length. The two chromatids of a chromosome are attached at a point called **primary constriction** or **centromere.** On the sides of centromere disc shaped structures called **kinetochores** are present.

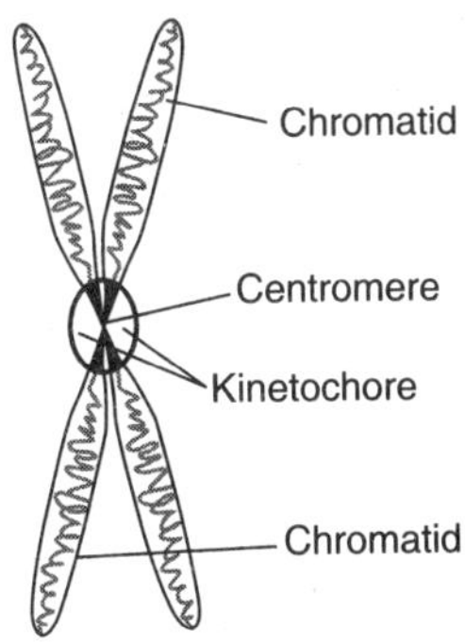

Fig. 2.22. Structure of a typical condensed chromosome.

Based on the position of centromere, chromosomes can be classified into four types (Fig. 2.23). (*i*) **Metacentric**—having median centromere with two equal arms of the chromosome (*ii*) **Sub-metacentric**—having sub-terminal centromere with one shorter arm and one longer arm (*iii*) **Acrocentric**—having centromere close to its end forming one extremely short and one very long arm. (*iv*) **Telocentric**—with terminal centromere.

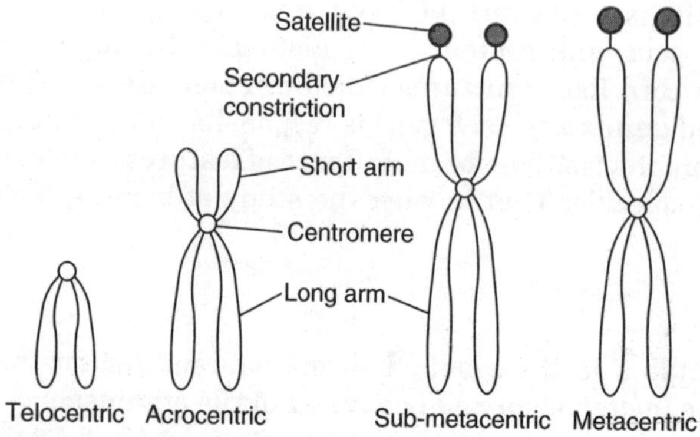

Fig. 2.23. Types of chromosomes based on the position of centromere.

Besides centromere, secondary constriction can also be observed in some chromosomes which if present in the distal region of an arm, would pinch off a small fragment called **satellite.**

Chromosome Number

The number of chromosomes is constant for a particular species. For example, there are 46 chromosomes in each body cell of human being ; 40 in mouse, 8 in *Drosophila* (fruit fly) ; 20 in maize ; 24 in tomato and 48 in potato. The body cells has a pair of chromosomes of each kind. This paired condition is called **diploid number** and is represented as **2n.** A set of unpaired chromosomes of each kind is called **haploid number,** and is represented as **n.** The gametes or sex cells (sperm or egg) contain haploid number (n) of chromosomes. Thus, each body cell of human being contains 46 or deploid ($2n$) number of chromosomes and gametes (sperm or ovum contains 23 or haploid (n) number of chromosomes. A haploid set of chromosome is called **genome.**

REVIEW EXERCISES

MULTIPLE CHOICE TYPE QUESTIONS

1. The term 'cell' was coined by
   (*a*) Robert Hooke (*b*) Robert Brown
   (*c*) Leeuwenhoek (*d*) M.J. Schleiden.
2. The term 'protoplasm' to the living substance present inside the cell was given by
   (*a*) Huxley (*b*) A. Flemming
   (*c*) J.E. Purkinje (*d*) R. Virchow.
3. The cell theory was proposed by
   (*a*) Watson and Crick (*b*) Beedle and Tatum
   (*c*) Darwin and Wallac (*d*) Schleiden and Schwann.
4. The barrier between the protoplasm and the outer environment in an animal cell is
   (*a*) Cell wall (*b*) Plasma membrane
   (*c*) Nuclear membrane (*d*) Tonoplast.
5. Ribosomes are the centres for
   (*a*) Respiration (*b*) Photosynthesis
   (*c*) Fat synthesis (*d*) Protein synthesis.

6. The site of aerobic respiration (oxidation) in a cell is
(*a*) Ribosomes (*b*) Golgi apparatus
(*c*) Chloroplasts (*d*) Mitochondria.

7. Which one of the following are called the suicide bags ?
(*a*) Centrosomes (*b*) Lysosomes
(*c*) Mesosomes (*d*) Chromosomes.

8. The membrane surrounding the vacuole of a plant cell is called
(*a*) Tonoplast (*b*) Plasma membrane
(*c*) Nuclear membrane (*d*) Cell wall.

9. Cell organelles found in plant cells only are
(*a*) Ribosomes (*b*) Endoplasmic reticulum
(*c*) Mitochondria (*d*) Plastids.

10. Lysosomes are the reservoirs of
(*a*) Fat (*b*) RNA
(*c*) Secretory glycoproteins (*d*) Hydrolytic enzymes.

OBJECTIVE TYPE QUESTIONS

1. Fill in the blanks with suitable words:
(*i*) The body of *Amoeba* is made up of .... cell.
(*ii*) The cell is the basic unit of .... and .... of the living beings.
(*iii*) The egg of .... is the largest cell.
(*iv*) Ribosomes are associated with the synthesis of ....
(*v*) The cellular organelles often referred as suicide bags are ....
(*vi*) The cell theory was given by .... and ....
(*vii*) Cell wall is found in .... cells.
(*viii*) Chromosomes are made up of .... and ....

2. Match the items of Coloumn I with those in Column II.

Column I	**Column II**
(*i*) Polysome	(*a*) Grana
(*ii*) Acid hydrolases	(*b*) Protein synthesis
(*iii*) SER	(*c*) Mitochondria
(*iv*) Thylakoid	(*d*) Lysosomes
(*v*) Cristae	(*e*) Steroid synthesis

3. Observe the relationship between the first two words and then fill in the suitable word at the fourth place:
(*i*) Centriole : 9 + 2 : : Cilia : ....
(*ii*) Chloroplast : photosynthesis : : amyloplast : ....
(*iii*) Mitochondria : light microscope : : ER : ....
(*iv*) Phagocytosis : cell eating : : pinocylosis : ....

TEST QUESTIONS

1. What is the significance of vacuole in a plant cell?
2. What does '*s*' refer to 70s and 80s ribosomes?
3. Mention the characteristics of prokaryotic cells.

4. Who proposed cell theory? Explain its main points.
5. What are the functions of cell wall?
6. List the functions of Golgi apparatus.
7. Describe the structure of a cilium/flagellum. Mention its main functions.
8. What are vacuoles? Name their types and functions.
9. Differentiate between prokaryotic and eukaryotic cells.
10. Describe the fluid mosaic model of plasma membrane.
11. Differentiate between the structures of a plant cell and an animal cell.
12. Describe the structure of Golgi apparatus and state its functions.
13. Comment on the cartwheel structure of centriole.
14. Differentiate between Rough Endoplasmic Reticulum (RER) and Smooth Endoplasmic Reticulum (SER).
15. Give the bio-chemical composition of plasma membrane. How are lipid molecules arranged in the membrane?
16. Briefly give the contribution of the following scientists in formulating the cell theory.
    (*i*) Schieden and Schwann
    (*ii*) Robert Virchaw
17. Eukaryotic cells have organelles which may
    (*i*) not be bounded by a membrane
    (*ii*) bound by a single membrane
    (*iii*) bound by a double membrane
    Group the various sub-cellular organelles into these three categories.
18. Justify the statement, "Mitochondria are power houses of the cell".
19. Are the different types of plastids interchangeable? If yes, give examples where they are getting converted from one type to another.
20. Describe different types of plastids and write their functions.
21. Describe the structure and functions of chloroplast.
22. Write the functions of the following :
    (*i*) Cell wall
    (*ii*) SER
    (*iii*) Centrioles
    (*iv*) Nucleolus.

3 Metabolic Engineering (Cellular Respiration)

All living organisms requires a continuous supply of energy for carrying out various functions such as uptake of materials, growth, development and movement. Energy is also required by some organisms for bioluminescence (light emission) and to keep their body warm. The energy needed by the cells is obtained through cellular respiration. Even the green plants that trap solar energy by photosynthesis and store it as organic compound depend on respiration for their continuous energy supply.

The concept of respiration had its beginning several millennia ago, when human beings first became aware that the act of breathing was most essential to life. The world 'respiration' came from the latin word 'respirare' which means to breath. *Cellular respiration is an energy releasing, enzymatically controlled, catabolic process which involves a stepwise oxidative breakdown of organic substances inside living cells.* Energy is not released in a single step in respiration nor it is directly used for cellular activities. Instead of energy is liberated in controlled fashion in several steps and is mostly stored in high energy bonds of ATP. About 100 kcal or 204 kJ of energy is released for each molecule of oxygen, used in respiration. This energy is then used to synthesise 6 molecules of ATP.

Cellular respiration resembles combustion (burning of coal, wood, oil etc.) in the (*i*) breakdown of complex organic substances, (*ii*) utilization of oxygen, (*iii*) production of carbon dioxide and (*iv*) release of energy, but there are some fundamental differences between the two processes. Combustion releases a large amount of energy in a single step, and most of it released as heat and sometimes partly as light. Cellular respiration releases energy in steps, and each step is coupled with the synthesis of ATP. Only a small amount of energy is dissipates as heat.

Differences between Respiration and Combustion

Respiration	*Combustion (Burning)*
1. Respiration is a biochemical process and occurs inside living cells.	1. Combustion is a physiochemical and non-cellular process.
2. It is under biological control.	2. It is an uncontrolled process.
3. Energy is released in stages as chemical bonds are broken in steps.	3. Energy is released in a single step as most of the chemical bonds break simultaneously.
4. Only a part of energy is lost as heat.	4. Most of the energy is liberated as heat.
5. Except in few cases, light is not emitted during respiration.	5. Light is often emitted during combustion.
6. Temperature is not allowed to rise.	6. Temperature becomes very high.
7. Most of the energy is trapped in ATP molecules.	7. No ATP is formed during combustion.
8. Each step of respiration is catalysed by an enzyme.	8. Enzymes are not involved in combustion.
9. A number of intermediates are formed for the synthesis of different organic compounds.	9. No intermediates are produced in combustion.
10. The oxidation of substrate occurs by the loss of hydrogen or electrons. Molecular oxygen is involved only in the last step (*i.e.*, terminal oxidation) of the process.	10. The substrate is directly oxidised in combustion.

ATP—The Energy Currency of the Cell

ATP was discovered by **Karl Lohmann** (1929). Its functioning through build up and hydrolysis of high energy phosphate bond was discovered by **Fritz Lipmann** (1941). Lipmann is called **'father of ATP cycle'.** In ATP, the last two phosphate radicals are attached by bonds of **high transfer potential.** They are also called **energy rich bonds** (Fig. 3.1) and are represented by the sign (~). The bond between second and third phosphate radicals possesses an energy equivalent of 8.9 Kcal/mole or 34 kJ/mole. While the bond linking the second phosphate radical with the first one has an energy equivalent of 6.5 Kcal/mole or 26 kJ/mole. The last phosphate can be very easily broken up and synthesised.

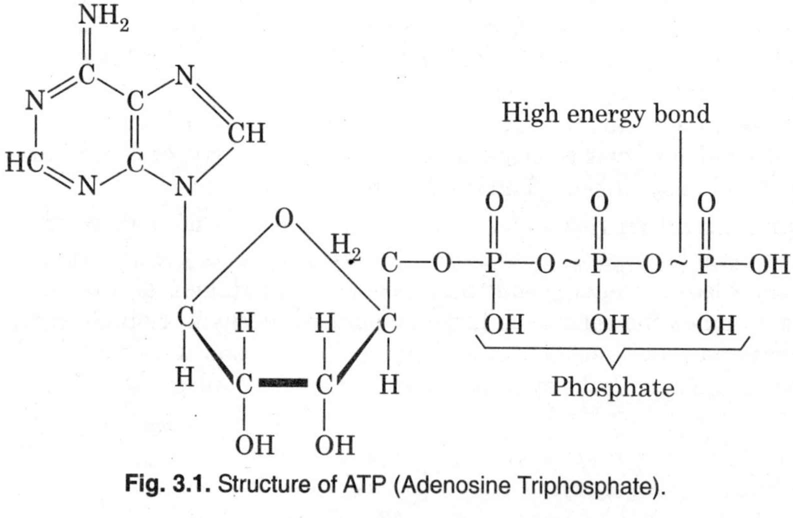

Fig. 3.1. Structure of ATP (Adenosine Triphosphate).

$$\text{Adenosine—P} \sim \text{P} \sim \text{P} + H_2O \xrightleftharpoons{\text{ATPase}} \text{Adenosine–P} \sim \text{P} + H_3PO_4 + \text{Energy.}$$

The easily available form of energy present in high energy bonds of ATP is called **biologically useful energy.** Hence, ATP functions as **energy currency** of the living cells. ATP can store small packets of energy as soon as the energy becomes available and minimise the wastage of energy. ATP molecules serve as energy pool of the cell, which ensures the availability of energy for various cellular functions.

EXCHANGE OF GASES

All cells have the necessary structural organisation and enzymes to carry out the oxidative process in small steps. In this process organic compounds, mainly carbohydrates, are oxidised by the removal of hydrogen and produce carbon dioxide. In most organisms, oxidation needs the participation of molecular oxygen obtained from outside. Oxygen is ultimately reduced to form water. Thus, oxidation of carbohydrates, besides releasing energy, leads to the formation of carbon dioxide and water.

The carbon dioxide formed in respiration must be removed from the cell. Therefore, respiration involves exchange of gases between the cell and its surroundings and also between the organism and its environment. An animal cell exchanges gases with the extracellular fluid while a plant cell does so with the air present in intercellular spaces. An organism shows exchange of gases in a liquid or gaseous environment depending upon the habitat. **Lavoisier** had demonstrated that animals take in oxygen from the air and give out carbon dioxide and water.

Some organisms, especially microorganisms do not use oxygen and some do not release carbon dioxide in respiration. Thus, gaseous exchange does not occur in these organisms.

RESPIRATORY SUBSTRATE

The organic substances which can be catabolised the living cells to release energy are called **respiratory substrates.**

- The most common respiratory substrate is glucose. It is formed from storage carbohydrates like starch in most plants and glycogen in animals and fungi.
- Fats may be used as respiratory substrate in some cases.
- Proteins are used in respiration rarely, as during germination of protein rich seeds and spores.
- In other organs, proteins are employed as respiratory substrate only when carbohydrates and fat reserves have been used up, as during long fasting.

However, fats and proteins are not directly used in respiration. Instead, they are initially broken down into intermediates identical with those in sugar oxidation. The intermediate metabolites from all three sources are finally oxidised through a common oxidative pathway, called the **Krebs or tricarboxylic acid cycle** (TCA cycle).

Respiration involving proteins as respiratory substrate is called **protoplasmic respiration,** whereas that uses carbohydrates or fat is termed as **floating respiration.** Protoplasmic respiration cannot be carried on for long as it liberates toxic ammonia.

TYPES OF RESPIRATION

Cellular respiration may be classified into two types, according to its dependence on oxygen—**aerobic** and **anaerobic.**

In aerobic respiration, organic food is completely oxidised with the help of oxygen (as terminal oxidant) into carbon dioxide and water. Aerobic respiration, occurs in mitochondria. 686 kcal or 2870 kJ of energy per molecules of glucose is liberated during aerobic respiration. The earlier estimate was 673 kcal.

$$C_6H_{12}O_6 + 6O_2 \xrightarrow{\text{Enzymes}} 6CO_2 + 6H_2O + \text{Energy (686 kcal/2870 kJ)}$$

In anaerobic respiration organic food is oxidised incompletely without oxygen being used as oxidant. Anaerobic respiration occurs in the cytoplasm and releases a small amount of energy. The common products of anaerobic respiration are carbon dioxide and ethyl alcohol or lactic acid. During anaerobic respiration, water is not produced.

$$C_6H_{12}O_6 \xrightarrow{\text{Enzymes}} 2CO_2 + \underset{\text{Ethyl alcohol}}{2C_2H_5OH} + \text{Energy (59 kcal/247 kJ)}$$

$$C_6H_{12}O_6 \xrightarrow{\text{Enzymes}} \underset{\text{Lactic acid}}{2C_3H_6O_3} + 36\ \text{kcal/150 kJ}$$

Anaerobic respiration, is the only mean of respiration in some micro-organisms (*e.g.*, yeast and some bacteria) and parasitic worms, In micro-organisms, the term **fermentation** is often used in place of anaerobic respiration. Although both possesses are defined as '*a complex series of cellular oxidation of organic compounds in which energy is made available to the cell in absence of oxygen*'. While fermentation can also occur outside the cell by extracellular enzymes, the anaerobic respiration occurs inside the cell.

In higher organisms, anaerobic respiration occurs as a temporary measure. It cannot continue for a long in these organisms because of (*i*) low yield of energy, (*ii*) decomposition of large amount of substrate so that little is left for growth and repair, (*iii*) toxic effect of end products in higher concentration and, (*iv*) inhibition of a number of physiological processes linked with aerobic respiration.

MECHANISM OF RESPIRATION

Both aerobic and anaerobic types of respiration have a common series of initial steps termed **glycolysis** (literally means sugar splitting). During glycolysis, carbohydrates are

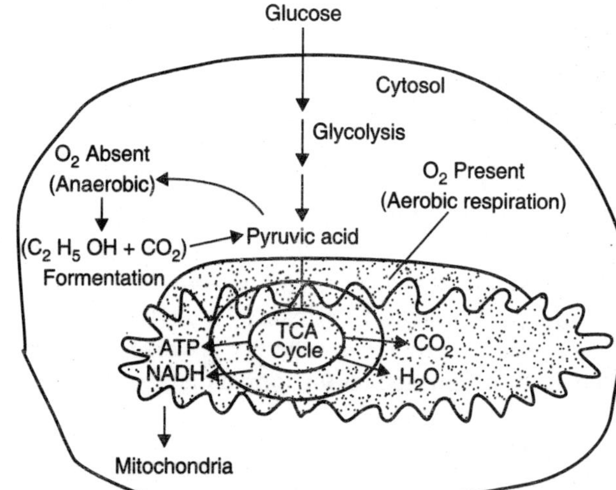

Fig. 3.2. The broad scheme of respiration.

converted into pyruvic acid through a series of enzymatic reactions. The pyruvic acid thus formed enters mitrochondria, where O_2 and the necessary enzymes are available, and pyruvic acid is finally converted into CO_2 and H_2O. This reaction series is known as **Krebs cycle** or **tricarboxylic cycle** (TCA cycle) or **citric acid cycle.** The general scheme of respiration is shown in (Fig. 3.2).

Glycolysis

It is also called **EMP pathway,** because it was discovered by three German scientists **Gustav Embden, Otto Meyerhof** and **J. Parnas.** Glycolysis is common to both aerobic and anaerobic modes of respiration and is therefore called **common pathway.** Glycolysis is the first stage in the breakdown of glucose and is common to all organisms. It occurs in the cytosol and results in the breakdown of glucose or similar hexose sugar into two molecules of a three carbon compound—pyruvic acid, releasing some energy (as ATP) and reducing power (as $NADH_2$). In plants, this glucose is derived from sucrose (which is the end product of photosynthetic carbon reactions) or from storage carbohydrates. Sucrose is converted into glucose and fructose by the enzyme **invertase,** and these two monosaccharides can readily enter the glycolytic pathway. The steps of glycolysis (Fig. 3.3) are described below.

1. **Phosphorylation.** Glucose phosphorylated to glucosc-6-phosphate by ATP in the presence of enzyme *hexokinase* and Mg^{2+}.

2. **Isomerisation.** Glucose-6-phosphate is changed into its isomer, fructose-6-phosphate with the help of enzyme *phosphohexose isomerase* (= *phosphogluco isomerase*).

3. **Phosphorylation.** Fructose-6-phosphate is phosphorylated by ATP to form fructose 1, 6-diphosphate in presence of enzyme *phospho-fructokinase* and Mg^{2+}.

Phosphorylation of glucose to fructose 1, 6-diphosphate activates the sugar and prevents it from getting out of the cell.

4. **Splitting.** Fructose 1, 6-diphosphate is then broken down into two molecules of triose phosphate (3 carbon sugars), *viz.* glyceraldehyde-3 phosphate (PGAL) and dehydroxyacetone phosphate (DiHAP). The reaction is catalysed by enzyme *eldolase.* Dihydroxy acetone phosphate is further converted into glyceraldehyde 3 phosphate with the help of enzyme *phospho triose isomerase.*

5. **Dehydrogenation and Phosphorylation.** Each glyceraldehyde 3-phosphate molecule loses hydrogen to NAD^+ to form NADH + H^+ and accept inorganic phosphate from phosphoric acid (H_3PO_4) to form 1, 3-diphosphoglycerate in the presence of enzyme *triose-phosphate dehydrogenase.*

6. **Dephosphorylation (ATP Formation).** One of the two phosphates of 1, 3-diphosphoglycerate is linked by high energy bond. In the presence of enzyme *phosphoglycerate kinase,* 1, 3-diphosphoglycerate is converted to 3-phosphoglycerate. One molecule of ATP is phosphorylated to ATP in the reaction. Formation of ATP directly from metabolites is known as **substrate level phosphorylation.**

7. **Isomerisation.** 3-phosphoglycerate is changed to its isomer 2-phosphoglycerate by the enzyme *phosphoglyceromutase.*

8. **Dehydration.** 2-phosphoglycerate loses a molecule of water in the presence of enzyme *enolase* and Mg^{2+}, and changes into phosphoenol pyruvate.

9. **Dephosphorylation (ATP Formation).** High energy phosphate group of phosphoenol pyruvate is transferred to a molecule of ADP with the help of the enzyme *pyruvate kinase* in the presence of Mg^{2+} and K^+. This produces pyruvate and a molecule of ATP by **substrate level phosphorylation.**

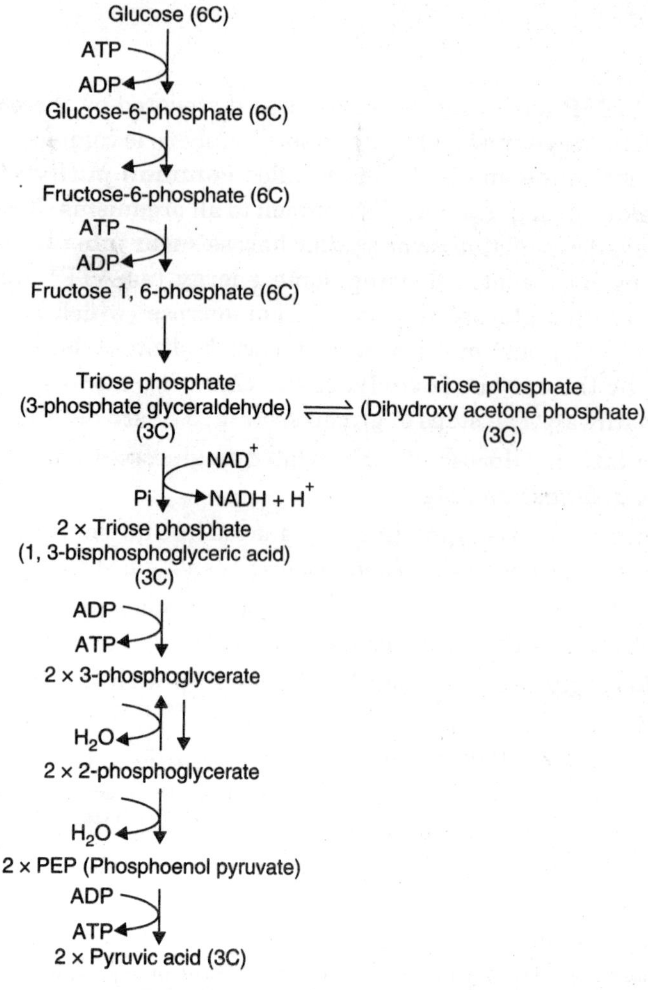

Fig. 3.3. Schematic representation of glycolysis.

Net Product of Glycolysis. In glycolysis two molecules of ATP are consumed during double phosphorylation of glucose to form fructose 1, 6-diphosphate. In return four molecules of ATP are produced by substrate level phosphorylation (conversion of 1, 3-diphosphoglycerate to 3-phosphoglycerate and phosphoenol pyruvate to pyruvate) and two molecules of $NADH_2$ are formed at the time of oxidation of glyceraldehyde 3-phosphate to 1, 3-diphosphoglycerate. The whole process may be expressed as under:

$$\text{Glucose} + 2NAD^+ + 2ADP + 2H_3PO_4 \longrightarrow 2\text{Pyruvate} + 2NADH + 2H^+ + 2ATP$$

Two molecules of $NADH + H^+$ on oxidation produce 6 molecules of ATP. Therefore, a net gain of 8 ATP molecules occurs during glycolysis.

Anaerobic Respiration (Anaerobic Oxidation of Pyruvate)

While studying alcoholic fermentation of glucose by yeast, **Pasteur** (1860) discovered that it can respire in the absence of molecular oxygen. This process was termed as **anaerobic respiration.**

Anaerobic respiration is an enzyme controlled partial breakdown of organic compounds (food) without using oxygen and releasing only a fraction of energy. Therefore, end products are never completely inorganic. It occurs in roots of some water logged plants, certain parasitic worms, animal muscles and some microorganisms *(e.g.,* moulds, bacteria). The term anaerobic respiration is often used in connection with higher organisms. In microorganisms, anaerobic respiration is often called **fermentation.** Fermentation is named after its product like **alcoholic fermentation, lactic acid fermentation. Buchner** (1897) found that fermentation could be caused by mixing sugar solution with yeast extract, instead of living yeast cells. The enzyme complex present in the extract was named as *zymase.* Because of the latter, fermentation is also called **zymosis.**

In anaerobic respiration electrons are removed from the substrate during oxidation but are not finally transferred to molecular oxygen. The final electron acceptors are frequently compounds such as pyruvic acid or acetaldehyde, which form a part of anaerobic respiratory pathway. The end products are lactic acid or ethyl alcohol and not water.

The mechanism of anaerobic respiration or fermentation resembles that of respiration upto glycolysis. Pyruvate formed at the end of glycolysis is anaerobically broken down to yield various products depending upon the organism and the type of tissue. The two common products are ethyl alcohol and lactic acid.

1. Alcoholic Fermentation

It occurs in fungi *(e.g., Rhizopus,* yeast) and bacteria. Yeast can respire both aerobically and anaerobically. Anaerobic respiration occurs in sugary solution causing fermentation. If the fungus is not in contact with atmosphere. Pyruvate first undergoes decaboxylation (removal of carboxyl group in the form of carbon dioxide) with the help of enzyme *pyruvate decarboxylase, transacetylase,* Mg^{2+} and TPP (thiamine pyrophosphate). This produces acetaldehyde and carbon dioxide from pyruvic acid.

$$\text{Pyruvate} \xrightarrow[\text{Mg}^{2+},\ \text{TPP}]{\text{Pyruvate decarboxylase}} \text{Acetaldehyde} + CO_2$$

Acetaldehyde then accepts hydrogen from $NADH_2$ and is reduced to ethyl alcohol (ethanol) producing oxidised NAD^+ (Fig. 3.4). The process is catalysed by the enzyme *alcohol dehydrogenase.*

$$\text{Acetaldehyde} + \text{NADH} + H^+ \xrightarrow{\text{Alcohol dehydrogenase}} \text{Alcohol} + NAD^+$$

Accumulation of ethanol by fermentation in a culture of yeast may stop further multiplication and lead to the death of cells. In the presence of oxygen, however, yeast can respire aerobically.

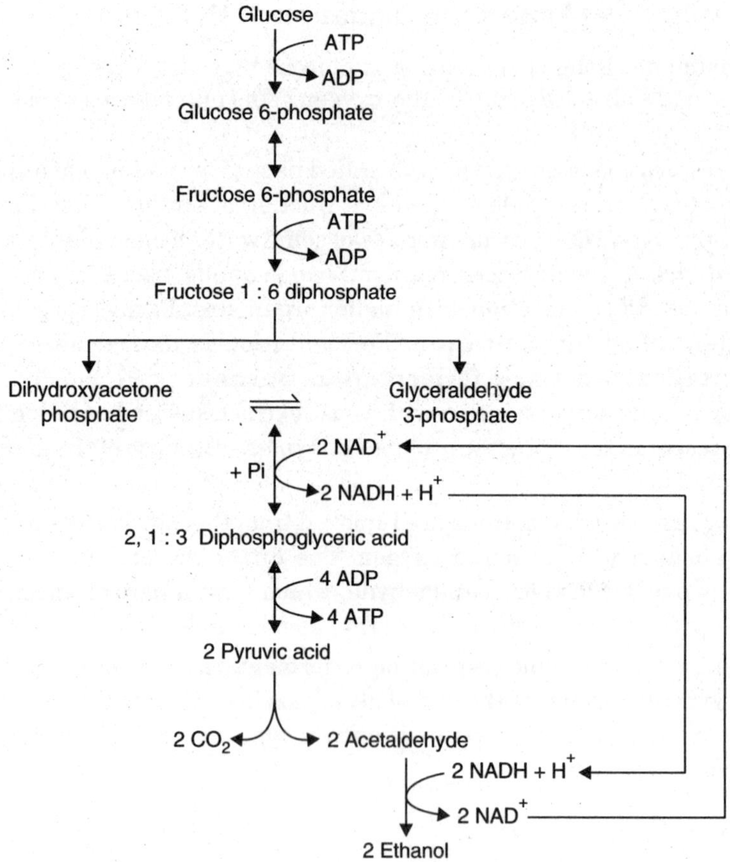

Fig. 3.4. Pathway of anaerobic respiration (fermentation) in yeast.

2. Lactic Acid Fermentation

It occurs in lactic acid bacteria (*e.g., Lactobacillus)* and muscles. Pyruvic acid produced in glycolysis is reduced by $NADH_2$ to form lactic acid. Carbon dioxide is not produced and $NADH_2$ is oxidised to NAD^+. The reaction is catalysed by the enzyme *lactic dehydrogenase* which requires FMN (Flavin mono nucleotide) and Zn^{2+}. NAD^+ is reutilised in glycolysis (Fig. 3.5).

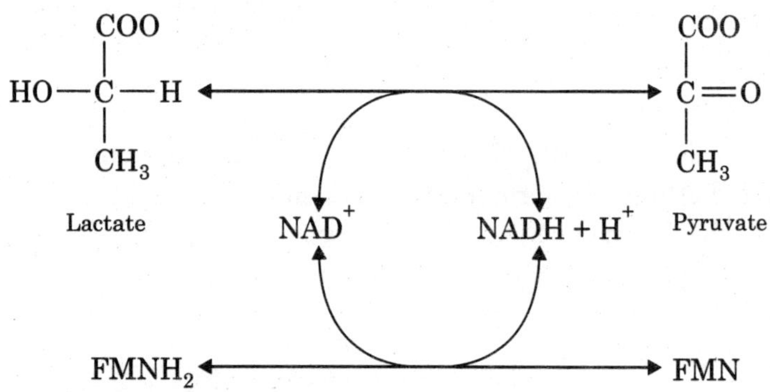

Fig. 3.5. Pathway of anaerobic respiration in muscle cells.

$$\text{Pyruvate} + \text{NADH} + \text{H}^+ \xrightarrow[\text{FMN, Zn}^{2+}]{\text{Lactic dehydrogenase}} \text{Lactic acid} + \text{NAD}^+$$

Skeletal muscles usually derive their energy by anaerobic respiration. After vigorous exercise, lactic acid accumulates, leading to muscular fatigue. During rest, however, the lactic acid is reconverted to pyruvic acid and is channeled back into the aerobic respiration pathway.

Anaerobic respiration yields much less energy than aerobic respiration. The main reasons are: (*i*) There is incomplete breakdown of respiratory substrate. (*ii*) One of the end product of anaerobic respiration is organic that still contains energy. (*iii*) $NADH_2$ produced during glycolysis is often reutilised. (*iv*) Regeneration of NAD^+ from $NADH_2$ does not produce ATP. (*v*) Electron transport chain is absent and (*vi*) Oxygen is not used for accepting electrons and protons.

During heavy physical work or strenuous exercise the demand for oxygen increases many fold. The oxygen does not diffuse into the skeletal muscle fibres at the required rate. Skeletal muscle fibres contain an oxygen storing pigment **myoglobin.** There are two types of muscle fibres—red (dark) and white (pale). Red muscle fibres are richer in myoglobin content than the white muscle fibres. Therefore, red fibres can utilise the oxygen stored in myoglobin to continue energy production over prolonged periods by aerobic oxidation of glucose. This enables them to perform sustained work over a long period. On the contrary, white fibres produce the energy needed for very fast and severe work by glycolysis, mainly because sufficient oxygen is not immediately available to them for such work. But white muscles accumulate lactic acid and get fatigued in a short time. Thus, athletes with a higher proportion of red fibres in their muscles are physiologically better equipped for sustained events such as cycling, running or swimming over long distances. Athletes with a higher proportion of white fibres in their muscles are equipped for fast, intense but short activities such as sprint, short put etc.

Importance of Anaerobic Respiration

(*i*) Anaerobic respiration is important during periods of oxygen deficiency, (*ii*) Various types of wines, beers and whisky are prepared through alcoholic fermentation of sugary solution with yeasts. (*iii*) Carbon dioxide, released by yeast cells in alcoholic fermentation is used in baking industry for making the bread light and spongy, (*iv*) Dairy industry depends upon the action of lactic acid bacteria which convert milk sugar, lactose to lactic acid. Lactic acid coagulates the milk protein casein and fuses droplets of milk fat. (*v*) Vinegar is obtained by the fermentation activity of acetic acid bacteria, (*vi*) Tea and tobacco leaves are cured (removal of their bitterness and development of characteristic flavour) by fermentation with certain bacteria. (*vii*) Fermentation is also employed in the preparation of ensilage (nutritive fodder for cattle), cleaning of hides, retting of stem fibres and production of industrial alcohols and organic acids, (*viii*) Decomposition dead bodies of organisms is carried out by fermentation, (*ix*) Fermenting organisms also spoil our food and may cause food poisoning.

Aerobic Respiration (Aerobic Oxidation of Pyruvate)

In aerobic respiration molecular oxygen acts as the ultimate acceptor of electrons and protons removed from the substrate. For this, the cell utilises the oxygen which diffuses into it from its surroundings. Aerobic respiration is carried out by the enzymes of the matrix and the inner membrane of the mitochondria. Carbon dioxide produced in the process diffuses out of the cell.

Pyruvic acid generated in the cytosol, is transported to mitochondria, and thus initiates the second phase of respiration. Before pyruvic acid enters Krebs cycle, operative in the mitochondria, one of the three carbon atoms of pyruvic acid is oxidised to carbon dioxide in a reaction called **oxidative decarboxylation.** *i.e.*, pyruvate is first decarboxylated, and then oxidised by the enzyme pyruvate dehydrogenase. The remaining 2-carbon acetate unit is readily accepted by a sulphur containing compound, coenzyme A (CoA), to form acetyl-CoA. It is the connecting link between glycolysis and Krebs cycle. During the process, NAD^+ is reduced to NADH. The reaction is summarised below :

$$\text{Pyruvic acid} + \text{CoA} \xrightarrow[\text{Mg}^{2+}]{\text{NAD}^+ \;\; \text{NADH}} \text{Acetyl CoA} + CO_2$$

During this process, two molecules of NADH are produced, *i.e.*, oxidative decarboxylation of two molecules of pyruvic acid produced during glycolysis. It results in a net gain of 6 ATP molecules (2 NADH × 3 = 6 ATP).

Beta oxidation of fatty acids also produces acetyl-CoA as the end product. Acetyl-CoA from both the sources acts as **substrate entrant** for Krebs cycle. It is also the **connecting link** between glycolysis and Krebs cycle.

Aerobic respiration consists of two distinct but mutually dependent processes :

1. **Krebs cycle** or **tricarboxylic acid cycle (TCA cycle)** in which reduced coenzymes are formed and carbon dioxide is evolved.

2. **Terminal oxidation** in which the reduced coenzymes are oxidised and oxygen is reduced to water.

KREB'S CYCLE OR TCA CYCLE

The cycle was discovered by British biochemist **Sir Hans Krebs** (1937). For the pioneer work Hans Krebs was awarded Nobel Prize in 1953. The cycle is also named as **Citric acid cycle** or **Tricarboxylic acid (TCA) cycle** after the initial product. Krebs cycle is stepwise oxidative and cyclic degradation of acetyl-CoA derived from pyruvate. Krebs cycle occurs in mitochondrial matrix and serves as a common oxidative pathway for carbohydrates, fats and proteins. (Figs. 3.6 and 3.7)

The actual citric acid cycle or Krebs cycle begins when acetyl-CoA enters into a reaction to form citric acid. It explains how pyruvate is broken down to CO_2 and water. It also highlighted the concept of cycles in metabolism.

The various steps of Krebs cycle (Fig. 3.8) are as follows :

1. **Condensation.** In the first reaction, 2-carbon compound acetyl CoA combines with 4-carbon compound oxaloacetate (OAA) in the presence of condensing enzyme *citrate synthetase* to form a tricarboxylic 6-carbon compound called citric acid and CoA is liberated. Citric acid is the first product of the Krebs cycle.

$$\text{Acetyl CoA} + \text{oxaloacetate} + H_2O \xrightarrow{\text{Citrate synthetase}} \text{Citrate} + \text{CoA}$$

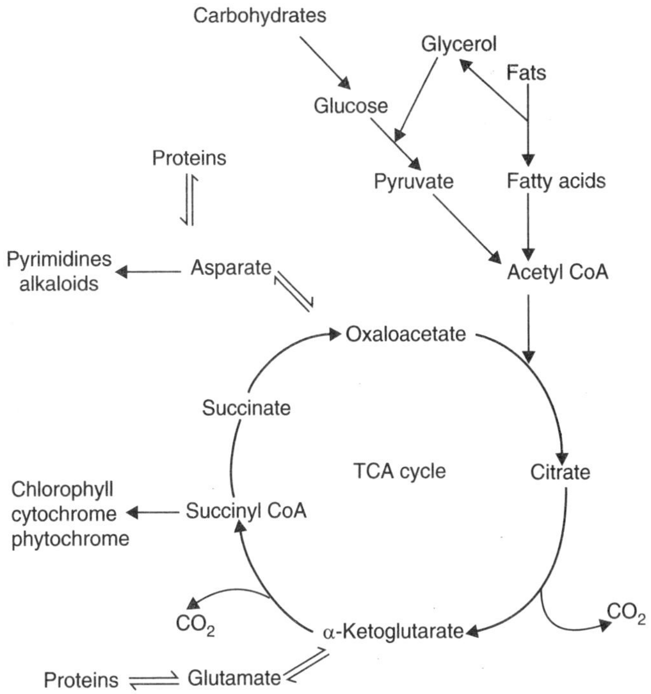

Fig. 3.6. TCA as common metabolic pathway of carbohydrates, fats and proteins.

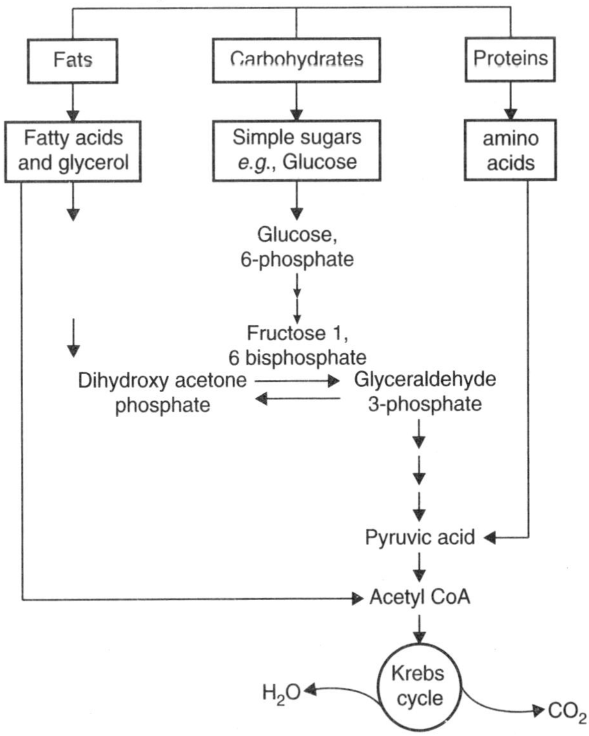

Fig. 3.7. Interrelationship among metabolic pathways showing respiration mediated breakdown of different organic molecules to CO_2 and H_2O.

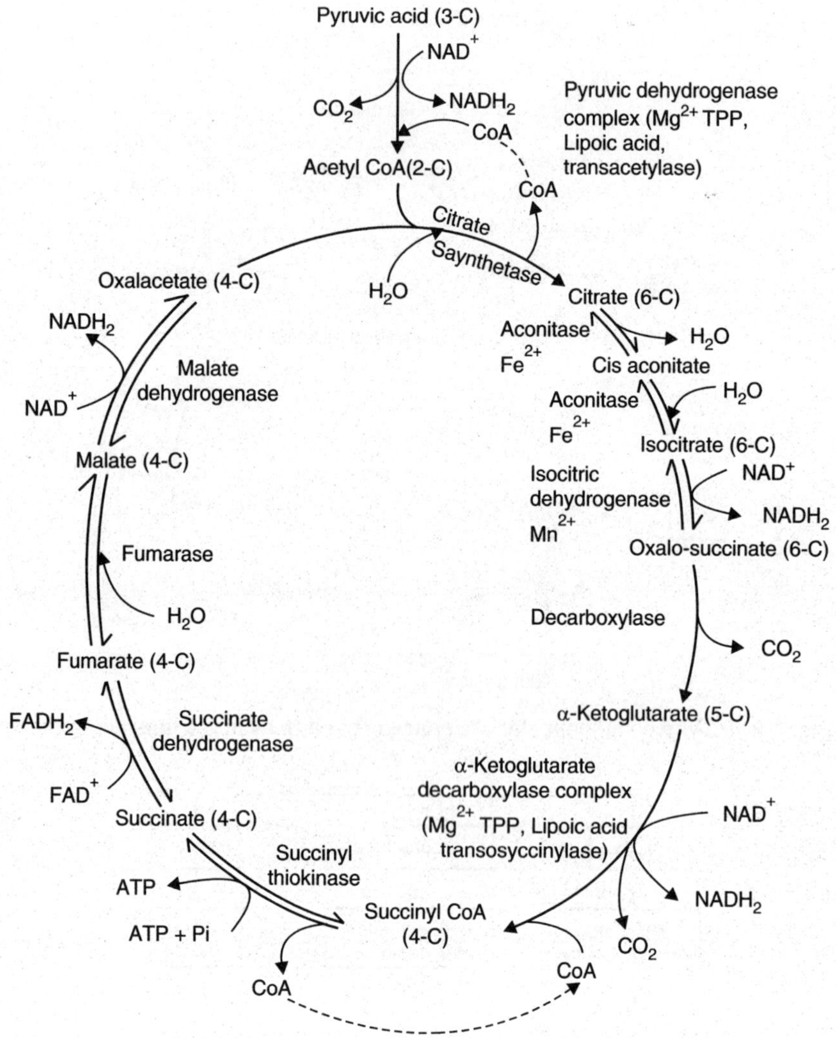

Fig. 3.8. Schematic representation of Krebs cycle or TCA cycle.

2. **Dehydration.** Citrate undergoes reorganisation in the presence of iron containing enzyme, *aconitase* forming 6-carbon *cis*-aconitate.

$$\text{Citrate} \xrightarrow{\text{Aconitase}} \textit{cis}\text{-aconitate} + H_2O$$

3. **Hydration.** *cis*-aconitate is further reorganised into 6-carbon isocitrate with the addition of water in the presence of enzyme *aconitase.*

$$\textit{Cis}\text{-aconitate} + H_2O \xrightarrow{\text{Aconitase}} \text{Isocitrate}$$

4. **Dehydrogenation.** Isocitrate is dehydrogenated to oxalosuccinate in the presence of enzyme *isocitrate dehydrogenase* and Mn^{2+}. A pair of hydrogen atoms is released which is accepted by NAD^+ to form $NADH + H^+$.

$$\text{Isocitrate} + NAD^+ \xrightarrow[Mn^{2+}]{\text{Isocitrate Dehydrogenase}} \text{Oxalosuccinate} + NADH + H^+$$

5. **Decarboxylation.** Oxalosuccinate is decarboxylated to form a 5-carbon α-ketoglutarate in the presence of enzyme *decarboxylase,* one molecule of carbon dioxide is released in the reaction.

$$\text{Oxalosuccinate} \xrightarrow{\text{Decarboxylase}} \alpha\text{-ketoglutarate} + CO_2$$

6. **Dehydrogenation and decarboxylation.** α-ketoglutarate is both dehydrogenated (with the help of NAD^+) and decarboxylated by an enzyme complex *α-ketoglutarate dehydrogenase.* The enzyme complex contains TPP and lipoic acid. The product combines with CoA to form a 4-carbon succinyl CoA.

$$\alpha\text{-ketoglutarate} + \text{CoA} + NAD^+ \xrightarrow[\text{TPP, Lipoic acid, } Mg^{2+}]{\alpha\text{-ketoglutarate Dehydrogenase}} \text{Succinyl CoA} + \text{NADH} + H^+ + CO_2$$

7. **Formation of GTP/ATP.** Succinyl CoA splits into succinate and CoA in the presence of enzyme *succinyl thiokinase.* The reaction releases sufficient energy to form GTP (in animals) and ATP (in plants).

$$\text{Succinyl CoA} + \text{GDP/GTP} + H_3PO_4 \xrightarrow{\text{Succinylthiokinase}} \text{Succinate} + \text{CoA} + \text{GTP/ATP}$$

8. **Dehydrogenation.** Succinate undergoes dehydrogenation to form 4-carbon fumarate with the help of enzyme *succinate dehydrogenase* and liberates a pair of hydrogen atom. The latter pass to FAD (Flavin adenine dinucleotide) to form $FADH_2$.

$$\text{Succinate} + \text{FAD} \xrightleftharpoons{\text{Succinate dehydrogenase}} \text{Fumarate} + FADH_2$$

9. **Hydration.** Fumarate is changed into 4-carbon malate with the addition of water in the presence of enzyme *fumarase.*

$$\text{Fumarate} + H_2O \xrightleftharpoons{\text{Fumarase}} \text{Malate}$$

10. **Dehydrogenation.** Malate is dehydrogenated in the presence of enzyme *malate dehydrogenase* to produce 4-carbon oxaloacetate. Hydrogen is accepted by $NAD^+/NADP^+$.

$$\text{Malate} + NAD^+/NADP^+ \xrightleftharpoons{\text{Malate dehydrogenase}} \text{Oxaloacetate} + \text{NADH/NADPH} + H^+$$

Oxaloacetate picks up another molecule of activated acetate to repeat the cycle.

The summary equation for this phase of respiration may be written as follows :

$$\text{Pyruvic acid} + 4NAD^+ + \text{FAD} + 2H_2O + \text{ADP} + \text{Pi} \longrightarrow 3CO_2 + 4\text{NADH} + 4H^+ + FADH_2 + \text{ATP}$$

During citric acid cycle, decarboxylation occurs at two steps releasing two molecules of CO_2. In this process 3 molecules of NAD^+ and one molecule of FAD (Flavin adenine dinucleotide) are reduced to produce 3-molecules of NADH and one molecule of $FADH_2$ respectively (Fig. 3.9). These reduced electron carriers pass on the hydrogen atoms to oxygen through electron transport system yielding ATP molecules. 3-molecules of NADH produce 9-molecules of ATP (3NADH × 3 = 9ATP), while the $FADH_2$ gives rise to 2-molecules of ATP ($1FADH_2 \times 2 = 2ATP$). In addition, one more ATP molecule is generated during the cycle. Thus, a total of 12 ATP molecules per molecule of pyruvic acid are produced. From each molecule of glucose, a total of 24-molecules of ATP are formed during the citric acid cycle.

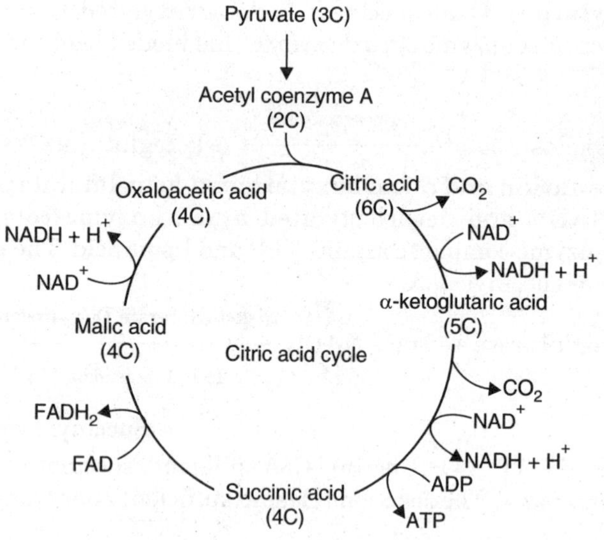

Fig. 3.9. The Citric acid cycle.

Differences between Aerobic and Fermentation Anaerobic Respiration

Aerobic Respiration	*Fermentation / Anaerobic Respiration*
1. Aerobic respiration is the normal mode of respiration of plants and animals.	1. Anaerobic respiration is the normal mode of respiration in some parasitic worms and micro-organisms. Others resort to it temporarily.
2. It uses oxygen for breaking the respiratory material into simple substances.	2. It does not use oxygen for breaking down the respiratory substrate.
3. Respiratory substrate is completely oxidised.	3. There is incomplete oxidation of respiratory substrate in anaerobic respiration.
4. It involves exchange of gases between the organism and the environment.	4. An exchange of gases is absent.
5. It yields inorganic end products.	5. At least one of the end products is organic. Inorganic substances may or may not be produced.
6. Aerobic respiration consists of these steps glycolysis, krebs cycle and terminal oxidation.	6. Anaerobic respiration consists of two steps-glycolysis and incomplete breakdown of pyruvate.
7. It always release carbon dioxide and water.	7. It may or may not release carbon dioxide. Water is usually not formed.
8. It occurs in cytoplasm and mitochondria.	8. It occurs in cytoplasm only.
9. Aerobic respiration involves electron transport.	9. Electron transport is absent in anaerobic respiration.
10. It produces 686 kcal of energy per gram mole of glucose.	10. It produces only 50 kcal of energy per gram mole of glucose.
11. It can occur indefinitely.	11. It cannot occur indefinitely (except in some microorganisms and parasitic worms) as it produces toxic substances and provides only a little energy.

Differences between Glycolysis and Krebs Cycle

Glycolysis	*Krebs Cycle / TCA Cycle*
1. It takes place in the cytoplasm.	1. It takes plate in the matrix of mitochondria.
2. It occurs both in aerobic and anaerobic respiration.	2. It occurs in aerobic respiration only.
3. It is a linear pathway.	3. It is a cyclic pathway.
4. Glycolysis is the first step of respiration in which glucose is broken down to the level of pyruvate.	4. Krebs cycle is the second step in respiration where an active acetyl group is broken down completely.
5. It degrades a molecule of glucose into 2 molecules of an organic substance pyruvate.	5. It degrades pyruvate completely into inorganic substances ($CO_2 + H_2O$).
6. It generates $2NADH_2$ and 2ATP for every molecule of glucose broken down.	6. It yields $6NADH_2$, $2FADH_2$ and 2ATP from 2 molecules of Acetyle CoA.
7. It does not produce CO_2.	7. CO_2 is evolved in Krebs cycle.
8. Oxygen is not required for glycolysis.	8. Krebs cycle uses oxygen as terminal oxidant.
9. It is not connected with oxidative phosphorylation.	9. Krebs cycle is connected with oxidative phosphorylation.

TERMINAL OXIDATION

It is the last step of aerobic respiration which involves the passage of both electrons and protons of reduced coenzymes to oxygen.

$$NADH_2/FADH_2 \rightleftharpoons NAD/FAD + 2H^+ + 2e^-$$

$$\frac{1}{2}O_2 + 2H^+ + 2e^- \longrightarrow H_2O$$

Terminal oxidation consists of two processes—electron transport and oxidative phosphorylation.

(*a*) **Electron Transport Chain.** The inner mitochondrial membrane contains some proteins which act as H^+ ions and electron transporting enzymes. The enzymes are arranged in ordered manner in a specific series called **Electron Transport Chain** (ETC) or **mitochondrial respiratory chain** or **Electron Transport System** (ETS). An electron transport chain is a series of enzymes and cytochromes in the inner mitochondrial membrane that take part in the passage of electrons from a substance to its ultimate acceptor. The electron carriers include flavins, iron-sulphur complexes, quinones and cytochromes. Most of them are prosthetic groups of proteins.

The glucose molecule is completely oxidised by the end of the citric acid cycle. But the energy is not released unless NADH and $FADH_2$ are oxidised through the electron transport system. Electrons from NADH produced in the mitochondrial matrix during citric acid cycle are oxidised by an NADH dehydrogenase **(Complex I)** and electrons are than transferred to ubiquinone located within the inner membrane. Ubiquinone also receives reducing equivalents *via.* $FADH_2$ that is produced during oxidation of succinate, through the activity of enzyme succinate dehydrogenase **(Complex II),** in the citric acid cycle. The reduced ubiquinone (ubiquinol) is than oxidised with the transfer of electrons to cytochrome C *via.* cytochrome bC_1 complex **(Complex III)** in a small protein attached to the outer surface of inner membrane and acts as a mobile carrier for transfer of electrons between complex III and IV, **Complex IV** refers to cytochrome *c* oxidise complex containing cytochrome *a* and a_3 and two copper centres (Fig. 3.10).

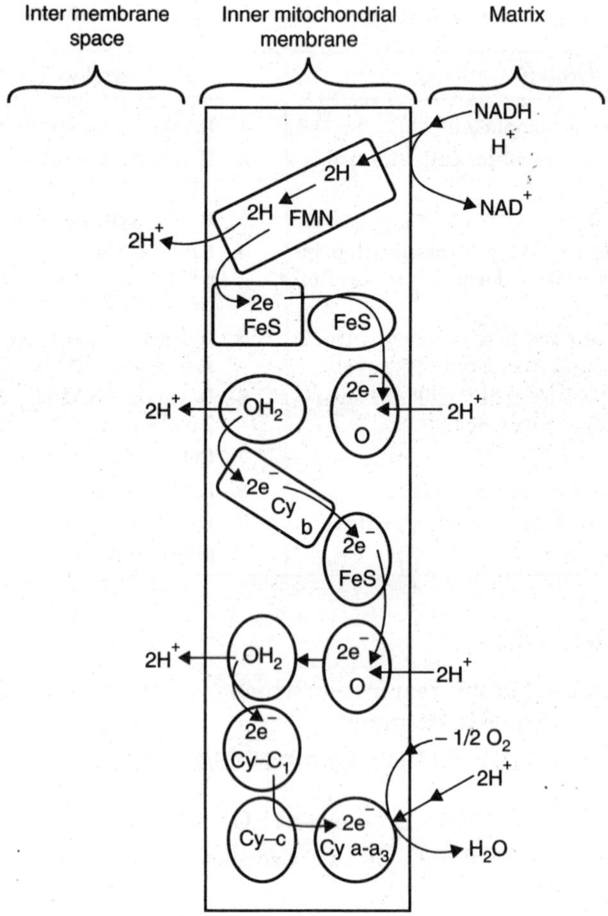

Fig. 3.10. Mitochondrial electron transport chain.

When the electrons pass from one carrier to another *via* complex I to IV in the electron transport chain, they are coupled to ATP synthase **(Complex V)** for the production of ATP from ADP and inorganic phosphate. The number of ATP molecules synthesised depends on nature of the electron donor. Oxidation of one molecule of NADH gives rise to 3 molecules of ATP, while that of one molecule of $FADH_2$, yields 2 molecules of ATP.

At each step of electron transport, the electron acceptor has a higher electron affinity than the electron donor. The energy from such electron transport is utilised in transporting protons (H^+) from the matrix across the inner membrane to its outerside (outer chamber). There are three such sites in the electron transport chain (Fig. 3.11). This creates a higher proton concentration outside the inner membrane (*i.e.,* in the outer chamber or inter mitochondrial space) than in the matrix. The difference in proton concentration on the outer and inner sides of the inner mitochondrial membrane is known as **proton gradient.**

(*b*) **Oxidative Phosphorylation.** Oxidative phosphorylation is the synthesis of energy rich ATP molecules with the help of energy liberated by oxidation of reduced coenzymes ($NADH_2$, $FADH_2$) produced during respiration. The enzyme required for their synthesis is called *ATP synthatase.* It is present in F_1 or head piece of F_0–F_1 or elementary particle. The particles are located in the inner mitochondrial membrane. The enzyme *ATP synthatase* becomes active in ATP formation only when there is proton gradient, having higher concentration of protons on the F_0 side (outer side) as compared to F_1 side (inner side).

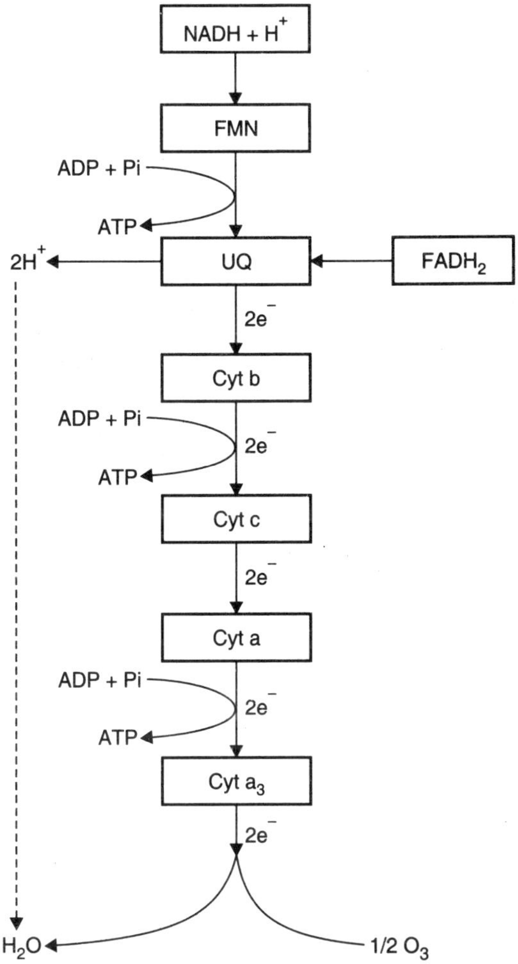

Fig. 3.11. A simplified system of terminal oxidation and oxidative phosphorylation.

Because of the higher proton concentration outside the inner membrane, protons return to the matrix down the proton gradient. Just as a flow of water from a higher to a lower level can be utilised to turn a water wheel or a hydroelectric turbine, the energy released by the flow of protons down the gradient is utilised in synthesising ATP. The enzyme *ATP synthetase* synthesises ATP from ADP and inorganic phosphate using the energy from the proton gradient (Chemiosmatic hypothesis of **Peter Mitchell,** 1961; Fig. 3.12)

Transport of two electrons from NADH + H^+ by the electron transport chain simultaneously transfers three pairs of protons to the outer compartment. One high energy ATP bond is produced per pair of protons returning to the matrix through the inner membrane particles. Therefore, oxidation of one molecule of $NADH_2$ produces 3 ATP molecules, while that of $FADH_2$ forms only 2 ATP molecules, as the latter donates its electron further down the chain.

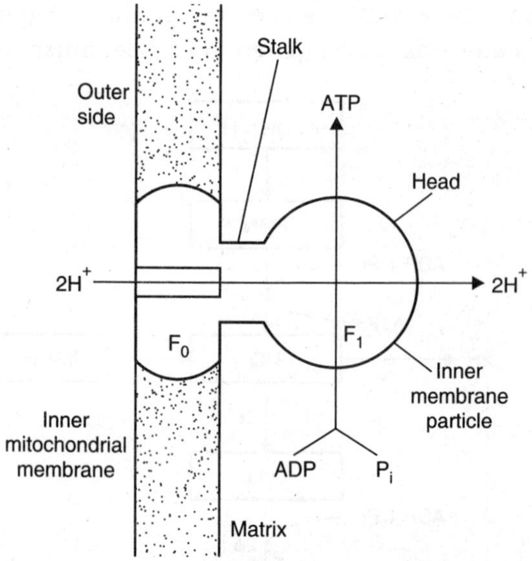

Fig. 3.12. ATP synthesis by F_0–F_1 particles or inner membrane particles of mitochondria.

Net gain of ATP. The details of ATP produced per molecule of glucose during aerobic respiration are given in Table 3.1.

Table 3.1. ATP Molecules Produced During Respiration

Stage of Respiration	*Source*	*Number of ATP Molecules Produced*
Glycolysis	Direct 2 molecules of NADH (One molecule of NADH yield 3 molecules of ATP)	2 6
Pyruvic acid of Acelyl-CoA	2 molecules of NADH	6
Citric acid cycle (Krebs cycle)	—6 NADH —2 $FADH_2$ ($FADH_2$ produces only 2 molecules of ATP) —Direct	18 4 2
Total yield of ATP molecules		38

There is a net gain of 38 ATP molecules during aerobic respiration of one molecule of glucose. In **eukaryotic** cells, $NADH_2$ produced in glycolysis are transferred to mitochondria for ATP synthesis. For this, two shuttle systems operate at the inner mitochondrial membrane.

(*i*) $NADH_2 \longrightarrow NAD \longrightarrow NADH_2$ (More efficient shuttle system), and

(*ii*) $NADH_2 \longrightarrow FAD \longrightarrow FADH_2$ (Less ellicient shuttle system)

The former operates in liver, heart and kidney cells. No energy is spent in this type of transfer. The second method occurs in muscle and nerve cells. It spent 2 ATP molecules in transporting 2 $NADH_2$ to the mitochondria, which lowers the energy level of 2 $NADH_2$ by 2 ATP molecules.

Thus, a total of 38 ATP molecules are produced per molecule of glucose oxidised in aerobic prokaryotes and, heart, liver and kidney cells of eukaryotes. Whereas, the net gain in most of the eukaryotic cells (*i.e.*, muscle and nerve cells) is 36 ATP molecules (as 2ATP molecules are consumed in transporting $NADH_2$ molecules into mitochondria). ATP molecules are transported out of mitochondria to cytoplasm through faciliated diffusion.

Efficiency of Aerobic Respiration

One molecule of ATP yields 34 kJ of energy. The energy yield from 38 ATP molecules comes to 1292 kJ.

The energy yield from complete oxidation of glucose is 2870 kJ. Thus, the efficiency of respiration is

$$\frac{1292}{2870} \times 100 = 45 \text{ per cent.}$$

It shows that only a part of this energy is used to make ATP and much of the energy generate during respiration is released in the form of neat.

Significance of Citric Acid Cycle

- Citric acid cycle is the major pathway for the synthesis of reduced coenzymes.
- It helps in controlled release of energy during aerobic respiration.
- It is a common pathway of oxidative breakdown of carbohydrates, fatty acids and amino acids. Amino acids entres the Krebs cycle directly as glutamate and aspartate after their deamination. Fatty acids undergo (β-oxidation to produce acetyl CoA which acts as substrata entrant for Krebs cycle.
- Acetyl CoA produces compounds for the synthesis of fatty acids, cutin, aromatic compounds, phytol chain of chlorophyll, carotenoids, terpenes and gibberellins.
- Succinyl CoA is involved in the synthesis of pyrrole compounds of chlorophyll, cytochrome and photochrome.
- The amino acid aspartate formed from oxaloacetate, takes part in the synthesis of pyrimidines and alkaloids.

Krebs cycle is a catabolic pathway, it also provides a number of intermediates for anabolic pathway. Therefore, Krebs cycle is often called **amphibolic pathway.**

Utility of Stepwise Oxidation

Stepwise metabolism serves several purposes:

- The stepwise release of chemical bond energy enables the cell to utilize a relatively higher proportion of that energy in ATP synthesis.
- It minimise the wastage of energy.
- It keeps the temperature of the cell low to prevent its burning.
- The activities of enzymes for different steps of respiration can be enhanced or inhibited by specific compounds. This helps in controlling the rate of respiration and the energy output according to the need of the cell.
- The intermediates of respiratory pathway can be used for the synthesis of other biomolecules such as amino acids.

REVIEW EXERCISES

MULTIPLE CHOICE TYPE QUESTIONS

1. The reactions of TCA cycle occur in
   (a) Grana of the chloroplast (b) Endoplasmic reticulum
   (c) Mitochondria (d) Cytoplasm.
2. In eukaryotes the complete oxidation of a molecules of glucose results in the net grain of
   (a) 2 molecules of ATP (b) 36 molecules of ATP
   (c) 4 molecules of ATP (d) 38 molecules of ATP.
3. Mitochondria are called power houses of the cells. Which of the following observations support this statement.
   (a) Mitochondria Synthesize ATP
   (b) Mitochondria have a double membrane
   (c) The enzymes of the krebs cycle and the cytochromes are found in mitochondria
   (d) Mitochondria are found in almost all plant and animal cells.
4. Lactic acid is formed by the process of
   (a) Fermentation (b) Krebs cycle
   (c) Glycolysis (d) Oxidative phosphorylation.
5. Krebs cycle beings with the reaction
   (a) OAA + Acetyl CoA (b) Citric acid + Acetyl CoA
   (c) OAA + Pyruvic acid (d) OAA + citric acid.
6. During which stage in the complete oxidation of glucose are the greatest number of ATP molecules formed from ADP?
   (a) Glycolysis (b) Krebs cycle
   (c) Electron transport chain (d) Conversion of pyruvic acid to acetyl CoA.
7. In glycolysis, during oxidation electrons are removed by
   (a) Glyceraldehydre 3 phosphate (b) NAD^+
   (c) Molecular oxygen (d) ATP.
8. The intermediate between glycolysis and TCA cycle is
   (a) Pyruvic acid (b) Glucose 1–6 diphosphate
   (c) Oxaloacetate (d) Acetyl CoA.
9. EMP can generate a total of
   (a) 6 ATP (b) 8 ATP
   (c) 24 ATP (d) 38 ATP.
10. The end product of aerobic respiration are
   (a) Sugar and oxygen (b) CO_2, H_2O and energy
   (c) Water and energy (d) CO_2 and energy.

OBJECTIVE TYPE QUESTIONS

1. Fill in the blanks with suitable words:
   (i) Glycolysis takes place in ______.
   (ii) Kreb cycle takes place in ______.
   (iii) $F_0 - F_1$ complex participates in the synthesis of ______.
   (iv) Acetyl CoA is formed from ______ and coenzyme A.
   (v) During glycolysis ______ molecules of NADH are formed from one molecule of glucose.

2. Match the items of Column I with appropriate terms of Column II.

Column I	Column II
(*a*) Glycolysis	(*i*) Terminal oxidation
(*b*) Krebs cycle	(*ii*) Ethanol and CO_2
(*c*) ETC	(*iii*) ATP
(*d*) Yeast	(*iv*) CO_2 and water
(*e*) $F_0 - F_1$ particles	(*v*) Pyruvate

TEST QUESTIONS

1. What are respiratory substrate? Name the most common respiratory substrate used in respiration.

2. Differentiate between:

(*i*) Respiration and combustion

(*ii*) Aerobic respiration and Fermentation

(*iii*) Glycolysis and Krebs cycle.

3. Where does glycolysis occur in a cell? Explain its different steps.

4. Explain the major steps in krebs cycle. Where does this occur in a cell?

5. Where is electron transport system operative in mitochondria? Explain the system highlighting the role of oxygen.

6. Calculate the efficiency of respiration in the living system.

7. Explain the main steps of aerobic respiration.

8. Illustrate the mechanism of electron transport system.

9. Enumerate the steps involved in the breakdown of carbohydrates upto pyruvic acid.

10. What are the advantages of stepwise oxidation of glucose during respiration?

11. Enumerate the significance of citric acid cycle in a living cell.

UNIT–2

MOLECULAR BIOLOGY, GENOMICS AND PROTEOMICS

4 DNA and RNA : The Molecular Basis of Heredity

The living matter is made up of different types of biomolecules. Therefore, genetic material should also be of biochemical nature. For a biochemical to be genetic material, it must have a high degree of variety and a mechanism of its replication. Nucleic acids possess both variety and mechanism of replication.

Nucleic acid was first isolated by a Swiss biochemist **Friedrick Miescher** (1869), from the nuclei of pus cells and named this substance as **nuclein.** The compound was later renamed as **nucleic acid,** due to its acidic properties by **Altman** (1899).

In a series of experiments conducted by **Frederick Griffith** (1928) on *Streptococcus pneumoniae* (bacterium responsible for pneumonia), and later by **Avery, Mac Leod** and **McCarty** (1944), it was concluded that DNA is the heredity material. The unequivocal proof that DNA is the genetic material came from the experiments of **Alfred Harshey** and **Martha Chase** (1952), who worked with viruses (bacteriophages) that infect bacteria.

Nucleic acids are of two types—**Deoxyribonucleic acid (DNA)** and **ribonucleic acid (RNA).** DNA acts as the genetic material in most of the organisms. RNA, though also acts as a genetic material in some viruses, mostly functions as a messenger, adapter, structural, and in certain cases as a catalytic molecules.

CHEMICAL COMPOSITION OF NUCLEIC ACIDS

Nucleic acids are the largest macromolecules found in the living world. They are the linear polymeres of purine and pyrimidine nucleotides. The nucleotides are compounds of carbon, hydrogen, oxygen, nitrogen and phosphorus. Each nucleotide is made up of three small chemicals—a pentose sugar, a nitrogenous base and a phosphate group (Fig. 4.1).

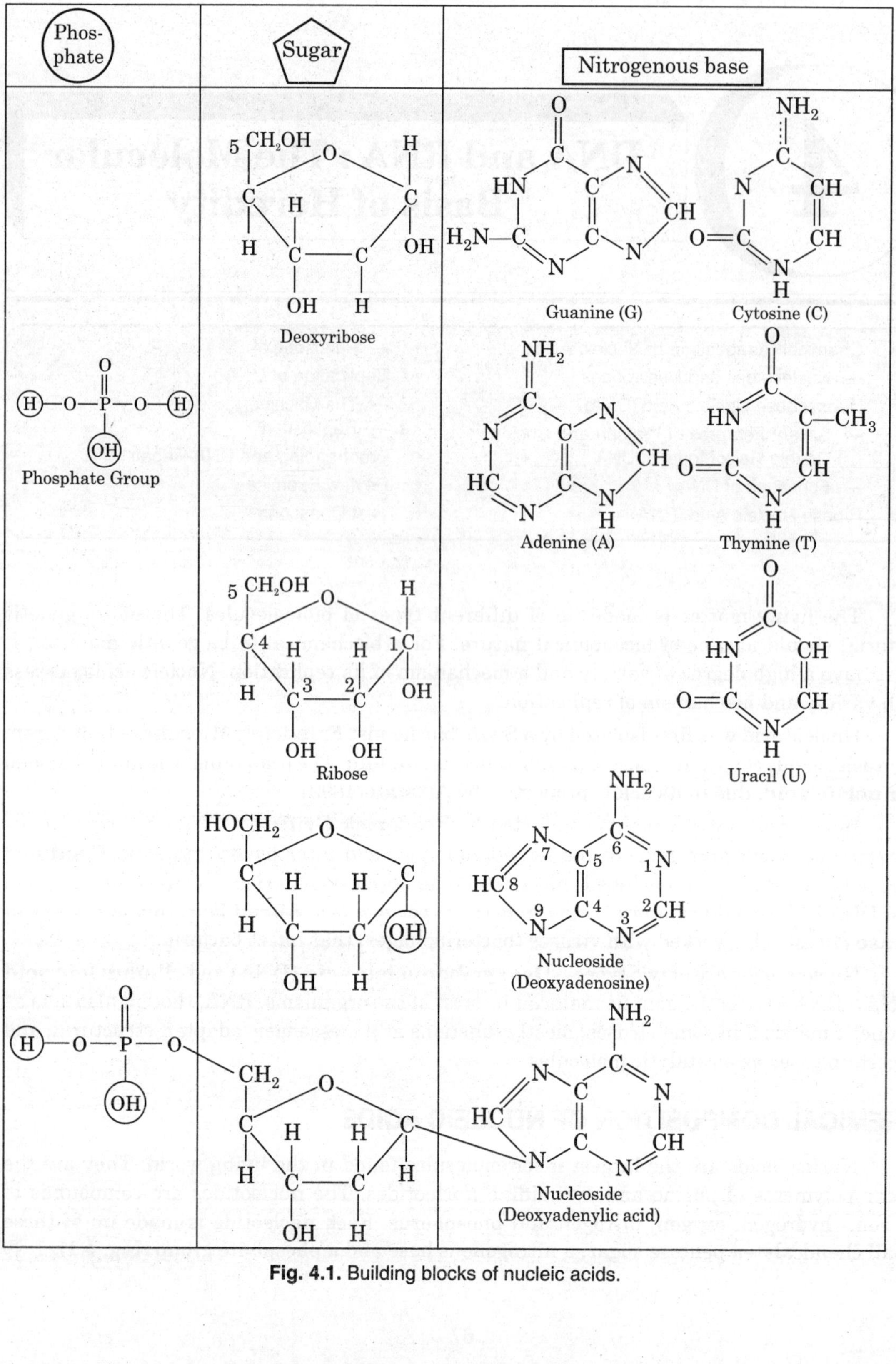

Fig. 4.1. Building blocks of nucleic acids.

1. **Pentose Sugar.** The sugar molecule in a nucleotide may be ribose ($C_5H_{10}O_5$) or deoxyribose ($C_5H_{10}O_4$). Both sugars are in furanose or pentagon state with one oxygen and four carbon atoms. The fifth carbon along with CH_2OH occurs outside the ring. Ribose molecule differs from deoxyribose molecule in having a hydroxyl (—OH) group instead of hydrogen at carbon 2 (Fig. 4.1). Deoxyribose sugar is found in the nucleotides forming DNA, while ribose occurs in the nucleotides forming RNA and a number of other compounds like AMP, ATP, NAD, FAD, etc.

2. **Nitrogenous Base.** The nitrogen bases occurring in nucleotides are heterocyclic compounds. They are of two types—**purines** and **pyrimidines.** Pyrimidines are six membered rings, while purines are nine membered double rings (Fig. 4.1). Purines are of two types—**Adenine (A)** and **Guanine (G)**, while pyrimidines are of three types—**Thymine (T), Cytosine (C)** and **Uracil (U).**

Differences between Purines and Pyrimidines

Purines	*Pyrimidines*
1. Purines are larger than pyrimidine molecules.	1. Pyrimidines are smaller than purine molecules.
2. They are nine membered molecules with double rings.	2. They are six membered molecules with single ring.
3. A purine contains four nitrogen atoms at 1′, 3′ and 9′ positions. **Examples.** Adenine (A) and Guanine (G).	3. A pyrimidine has nitrogen atoms at two places, 1′ and 3′ positions. **Examples.** Thymine (T), Cytosine (C) and Uracil (U).

3. **Phosphate Group (Phosphoric Acid).** It is a triportic acid with a centrally placed phosphorus atom, which is attached with three hydroxyl (—OH) group and one oxygen (—O) atom.

Nucleosides and Nucleotides

Nucleosides

The nitrogen base joins to the sugar molecule at its carbon atom 1′ by a **glycosidic bond** by one of its nitrogen atoms (usually 3 in pyrimidines and 9 in purines). A combination of the two is called **nucleoside.** For example, a combination of adenine-ribose or adenine-deoxyribose are nucleosides. A nucleoside with ribose sugar is known as **ribonucleoside** or simply **riboside** and a nucleoside with deoxyribose sugar is termed as **deoxyribonucleoside** or simply **deoxyriboside.** Thymine forms nucleoside with only deoxyribose sugar, uracil, similarly, produces nucleoside with ribose sugar. Other nitrogen bases form nucleoside with both ribose and deoxyribose sugars. The various nucleosides are **adenosine** (adenine + ribose), **deoxyadenosine** (adenine + deoxyribose), **guanosine** (guanine + ribose), **deoxyguanosine** (guanine + deoxyribose), **cytidine** (cytosine + ribose), **deoxycytidine** (cytosine + deoxyribosome), **deoxythymidine** (thymine + deoxyribose) and **uradine** (uracil + ribose).

Nucleotides are phosphoric acid esters of nucleosides. Phosphate combines with the sugar of the nucleoside at its 5′ carbon atom by an ester bond to form a **nucleotide** or **nucleoside monophosphate** (Fig. 4.1). Both glycosidic and ester bonds are formed by condensation reactions involving elimination of water. A nucleotide having ribose sugar is called **ribonucleotide** or simply **ribotide,** and a nucleotide containing deoxyribose is termed as **deoxyribonucletide** or simply **deoxyribotide.** The different nucleotides are adenylic acid or adenosine monophosphate (AMP), deoxyadenylic acid or deoxyadenosine monophosphate (dAMP), guanylic acid or guanosine monophosphate (GMP), deoxyguanylic acid or deoxyguanosine monophosphate (dGMP). Cytidylic acid or cytidine monophosphate (CMP), deoxycitidylic acid or deoxycytidine monophosphate (dCMP), deoxythymidylic acid or deoxythymidine monophosphate (dTMP), and uridylic acid or uridine monophosphate (UMP) (Table 4.1). The free occurring nucleotides may have upto three phosphate groups.

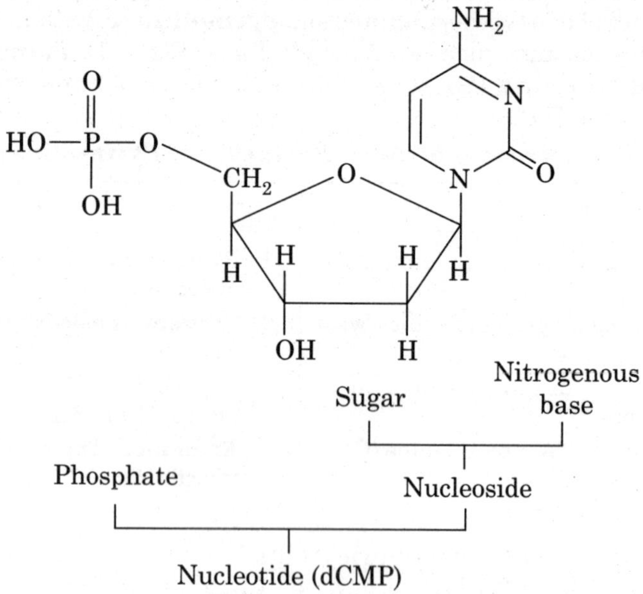

Fig. 4.2. Formation of a nucleoside and a nucleotide.

Table 4.1. Different Types of Nucleotides Occurring in DNA and RNA

Ribonucleotide	*Composition*	*Symbol*	*Deoxyribonucletide*	*Composition*	*Symbol*
Adenosine monophosphate (Adenylic acid) (or Adenylate)	Ribose + Adenine + Phosphate	AMP	Deoxyadenosine monophosphate (Deoxyadenylic acid)	Deoxyribose + Adenine + Phosphate	dAMP
Guanosine monophosphate (Guanylic acid or Guanylate)	Ribose + Guanine + Phosphate	GMP	Deoxyguanosine monophosphate (Deoxyguanylic acid)	Deoxyribose + Guanine + Phosphate	dGMP
Cytidine monophosphate (Cytidylic acid)	Ribose + Cytosine + Phosphate	CMP	Deoxycytidine monophosphate	Deoxyribose + Cytosine + Phosphate	dCMP
Uridine monophosphate (Uridylic acid or Uridylate)	Ribose + Uracil + Phosphate	UMP	Deoxythymdine monophosphate	Deoxyribose + Thymine + Phosphate	dTMP

Differences between Nucleoside and Nucleotide

Nucleoside	*Nucleotide*
1. Nucleoside is a compound formed by the combination of a nitrogen base and a pentose sugar. 2. It is slightly basic in nature. 3. A nucleoside is a component of a nucleotide.	1. Nucleotide is a compound formed by the combination of a nitrogen base, a pentose sugar and phosphate. 2. It is acidic in nature. 3. A nucleotide may be a component of nucleic acids, energy carriers or coenzymes.

DEOXYRIBOSE NUCLEIC ACID (DNA)

Occurrence. In prokaryotic cells, DNA lies coiled in the cytoplasm. In eukaryotic cells, DNA occurs mainly in the nucleus. It is called **nuclear DNA.** The nuclear DNA is linear and is associated with basic proteins (histones) to form deoxyribonucleo-protein (DNP). A small amount of DNA occur in mitochondria and plastids. It is called **extranuclear** or **organeller DNA.** The DNA found in prokaryotes, mitochondria and plastids is generally circular. It is not covered with proteins and is therefore, called **naked.** DNA is also present in some viruses.

Structure. DNA is a long chain polymer of over a million deoxyribonucleotides (*e.g.* 4.6×10^6 nucleotide pairs or base pairs (bp) in *Escherichia coli*, 6.6×10^9 nucleotide pairs (bp) in humans. It has two complementary unbranched strands of deoxyribonucleotide units. The two strands are spirally coiled around a common axis to form a **double helix** (some viruses are exception to have single stranded DNA). The double helix is considered to be right handed as proposed by **Watson** and **Crick** (1963), who discovered the structure of DNA.

The nucleotide of DNA or deoxyribonucleotide is formed by cross linkage of three substances—phosphoric acid, deoxyribose sugar and a nitrogen base. Four types of nitrogen bases are present in DNA. Two of them are purines—adenine (A) and guanine (G). The other two are pyrimidines—thymine (T) and cytosine (C). Thus, there are four types of deoxyribonucleotides in DNA. These are deoxyadenosine monophosphate (dAMP), deoxyguanosine monophosphate (dGMP), deoxy-thymidine monophosphate (dTMP) and deoxycytidine monophosphate (dCMP). The backbone of DNA chain or strand is made up of alternate deoxyribose and phosphoric acid groups. The phosphate group is connected to 5′ C of sugar residue of its own nucleotide and 3′ C of the sugar residue of the next nucleotide by phosphodiester bonds [Fig. 4.3 (*a*)]. At one end of DNA strand, the deoxyribose has its 3′ C free, while at the other end 5′ C of the last deoxyribose is free. They are respectively called 3′ and 5′ ends. Nitrogen bases are attached to 1′ C of the sugar by glycosidic bonds and lie at right angles to the longitudinal axis of DNA chains. A pyrimidine is attached to deoxyribose by its N-atom at 3′ position, while a purine joins by its N-atom at 9′ position.

The two chains of DNA are antiparallel, that is, they run parallel but in opposite directions. One chain runs in 5′–3′ direction and the other runs in 3′–5′ direction. The two chains are held together by hydrogen bonds between their bases. The purines of one chain pair with pyrimidines of the other. The pairing of purines and pyrimidines is also specific. Adenine (A), a purine is always paired with thymine (T), a pyrimidine by two hydrogen bonds, and guanine (G), a purine always paired with cytosine (C) a pyrimidine by three hydrogen bonds. This allows a sort of lock and key arrangement between large sized purine and small sized pyrimidine. The three hydrogen bond between cytosine and guanine (C≡G) occur at positions 1′–1′, 2′–6′ and 6′–2′, whereas two hydrogen bonds between adenine and thymine (A = T) occur at positions 1′–3′ and 6′–4′. The hydrogen bonds occur between the hydrogen of one base and the oxygen or nitrogen of the other base.

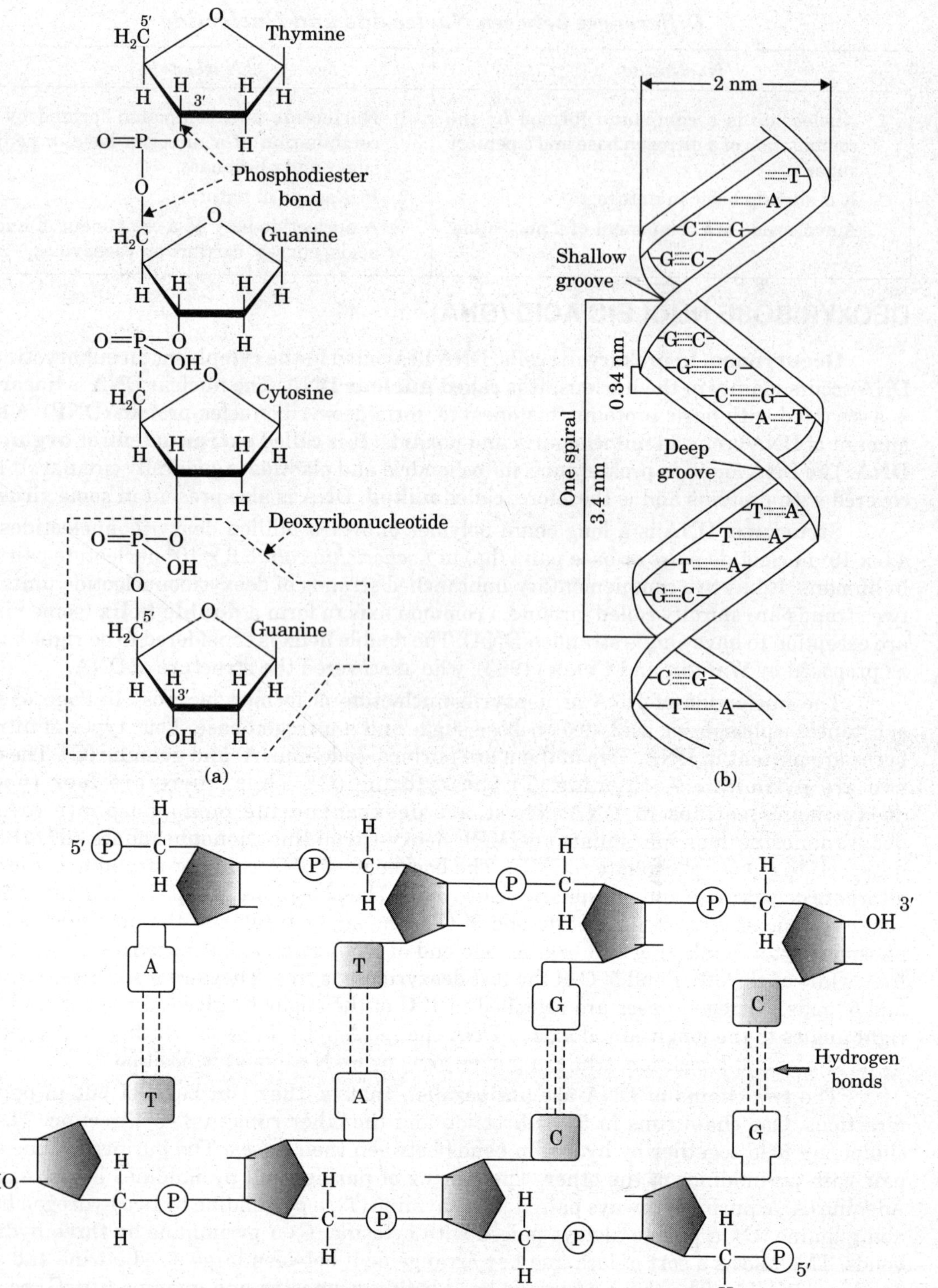

Fig. 4.3. (*a*) A polynucleotide strand of DNA; (*b*) Watson-Crick model of DNA double helix and (*c*) Double stranded polynucleotide chain of DNA.

Base pairs are complementary nitrogen bases present on the two DNA strands with a purine lying opposite a pyrimidine or *vice versa* [Fig. 4.3 (*b*) and (*c*)]. The larger sized purine always lies opposite to smaller sized pyrimidine. This provides an uniform thickness of 2 nm to the double helic DNA molecule.

One turn of the twisted double helix of DNA measures 3.4 nm, and consists of a stack of 10 nucleotides pairs. Thus, the adjacent nucleotides occur at a distance of 0.34 nm. At each base pair of ascent, the strand turns 36°. DNA molecule has a diameter of 2.0 nm. This diameter is sufficient to accommodate one purine and one pyrimidine nucleotide opposite to each other. The double helix has two types of alternate grooves : major or deep groove (2.2 nm wide) and minor or shallow groove (1.2 nm wide).

Although, only four types of base pairs are involved in the formation of DNA molecule, these base pairs may occur in any sequence and in any number of times. This gives rise an infinite variety of DNA molecules. However, each individual has a specific sequence of base pairs in its DNA.

Salient Features of Watson and Crick's Double Helix Model of DNA

- DNA is made of two polynucleotide chains, where the back bone is formed by sugar-phosphate, and base project inside.
- The two chains have anti-parallel polarity. It means, if one chain has the polarity $5' \rightarrow 3'$, the other has $3' \rightarrow 5'$.
- The bases in two strands are paired through hydrogen bond (H-bonds) forming base pairs (bp). Adenine forms two hydrogen bonds with thymine from opposite strand and vice versa. Similarly, Guanine is bonded with cytosine with three hydrogen bonds. As a result always a purine comes opposite to a pyrimidine. This generates approximately uniform distance between the two strands of the helix.
- The two chains are coiled in a right handed fashion. The pitch of the helix is 3.4 nm, and there are roughly 10 bp in each turn. Consequently, the distance between a bp in a helix is approximately equal to 0.34 nm.
- The plane of one base pair stacks over the other in double helix. This in addition to H-bonds, confers stability to the helical structure.

Functions of DNA

DNA performs a variety of functions in the living organisms:

- DNA is a genetic material which carries all the hereditary information. The latter is coded in the sequence of its nitrogen bases.
- Through the unique property of replication, DNA transfers the genetic information from one cell to its daughter and from one generation to the next.
- It gives rise to RNAs through the process of transcription.
- It controls the cell metabolism by directing the synthesis of specific proteins, enzymes and hormones, with the help of specific RNAs.
- It contributes to the evolution of the organisms by undergoing mutation *i.e.*, changes in the sequence of nitrogen bases due to deletion, addition or wrong replication.
- It also contributes in the formation of new combination of genes or recombination through crossing over during meiosis.
- It controls the differentiation and development of the organisms through differential functioning of its specific regions or genes.

RIBOSE NUCLEIC ACID (RNA)

Ribose nucleic acid is a single strand or chain which is formed by end to end polymerisation of a number of ribonucleotides. The polynucleotide is characterised by the presence of ribose sugar and uracil instead of thymine. A strand of RNA may contain 7–12,000 ribonucleotides. A ribonucleotides is formed of ribose sugar, phosphoric acid and a nitrogen base. The bases in RNA are Adenine (A), Guanine (G), Uracil (U) and Cytosine (C). The main strand is formed of alternate residues of phosphate and ribose sugar. Phosphate is attached to 5′ C of its sugar and 3′ C of next sugar (phosphodiester bond), similar to the arrangement found in DNA strand. Nitrogen bases are attached to sugars at 1′ C of the latter (Fig. 4.4).

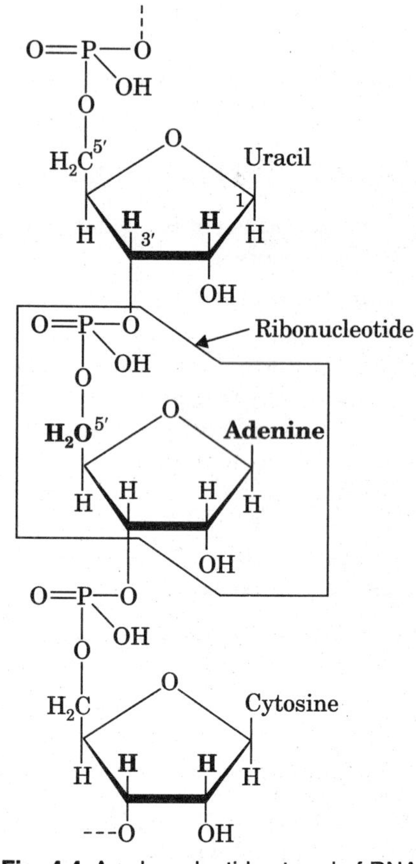

Fig. 4.4. A polynucleotide strand of RNA.

RNA is generally involved in protein synthesis, but in some viruses it also serves as genetic material. Thus, there are two types of RNA—genetic RNA and non-genetic RNA. **Genetic RNA** is the hereditary material in some viruses called riboviruses. The genetic RNA may be single stranded (*e.g.*, Tobacco Mosaic Virus) or double stranded (*e.g.*, Rheovirus). **Non-Genetic RNA** is present in all the cells, where DNA is the genetic material. There are three types of non-genetic RNA—ribosomal RNA (rRNA), transfer RNA (tRNA) and messenger RNA (mRNA). The three types of RNAs are transcribed from different regions of DNA template. Therefore, the sequence of nitrogen bases found on RNA strand is complementary to their sequence on DNA template. All the three types of non-genetic RNAs are synthesized in the nucleus but moves out into the cytoplasm. They play significant role in protein synthesis.

1. **Ribosomal RNA** (rRNA). It is the constituent of ribosomes. It form 70–80% of the total RNA content of the cell. Ribosomal RNA lies coiled in between and over the protein molecules of the ribosome. It helps tRNA and mRNA in the synthesis of proteins over the ribosomes.

2. **Transfer RNA (tRNA).** It is also called soluble RNA or sRNA. Transfer RNA constitute about 15% of the total RNA. tRNA is the smallest RNA consisting of 70–85 nucleotides. The nitrogen bases of some nucleotides get modified. This causes coiling of otherwise single stranded tRNA to form a **clover leaf** (in two dimension) or L-shaped (in three dimension) configuration (Fig. 4.5). tRNA has a site for binding amino acid. It lies at 3 end and has CCA—OH group. Opposite to the amino acid or AA binding site, there is an **anticodon** made up of three nitrogen bases for recognising and attaching to mRNA. There are two loops, one for attaching to ribosome (T ψ C, loop) and the second for binding to an enzyme (DHU or dihydroxyuridine loop). tRNA act as **adaptor molecules** which is meant for transfering amino acids to ribosome for synthesis of polypeptide. There are different tRNA for different amino acids. Some amino acids can however, be picked up by 2–6 tRNAs. Thus, there are over 100 types of tRNA in a cell.

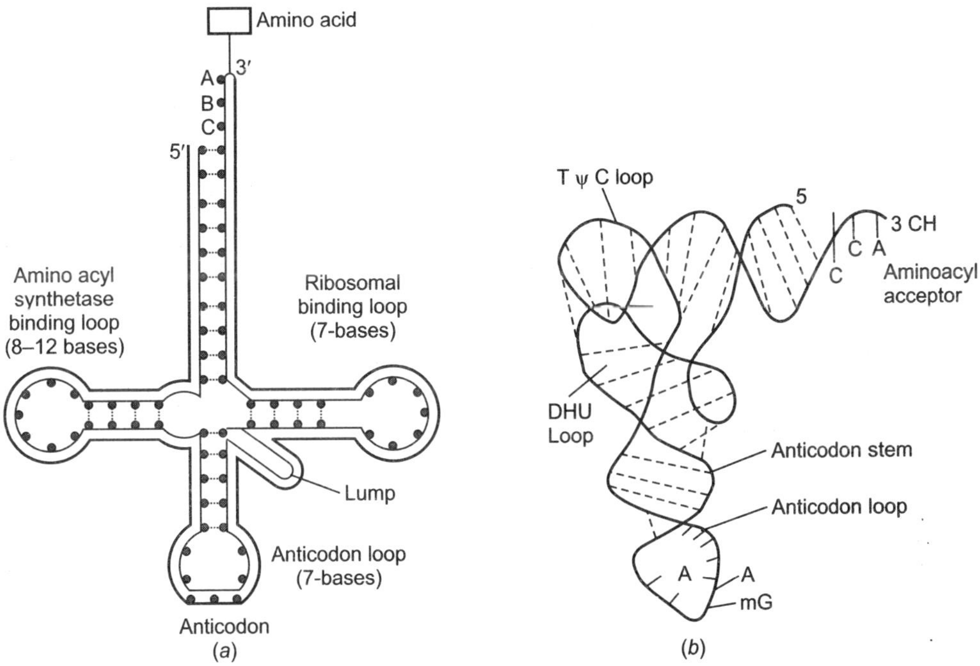

Fig. 4.5. Structure of tRNA; (*a*). Clover leaf (in two dimension) and (*b*) L-shaped (in three dimension) configuration.

3. **Messenger RNA (mRNA).** Messenger RNA is the longest of all RNAs. It constitute 2–5% of the total RNA content of the cell, mRNA carries genetic information from DNA for the formation of particular polypeptide. The information is present in the base sequence of its nucleotides. It is called **genetic code.** The sequence of nitrogen bases on mRNA form codons to specify amino acids in polypeptide chain. The 5′ terminus of mRNA has a methylated region, made up of an unusual nucleotide, methyl guanosine triphosphate ($^{m}G_{ppp}$), which acts as **cap** for attachment with ribosome. An initiator or start codon (AUG) follows the cap either immediately or after a small non-coding region. Then, there is coding region. It is followed by

terminator codon (UAA, UAG or UGA) a small non-coding region and poly A (consisting of 200–300 adenylate residues called **tail**) area at the 3′ terminus (Fig. 4.6). One mRNA molecule normally specifies a single polypeptide chain and said to be **monocistronic.** In some cases, mostly in prokaryotes one mRNA can specify more than one polypeptide and is described as **polycistronic.**

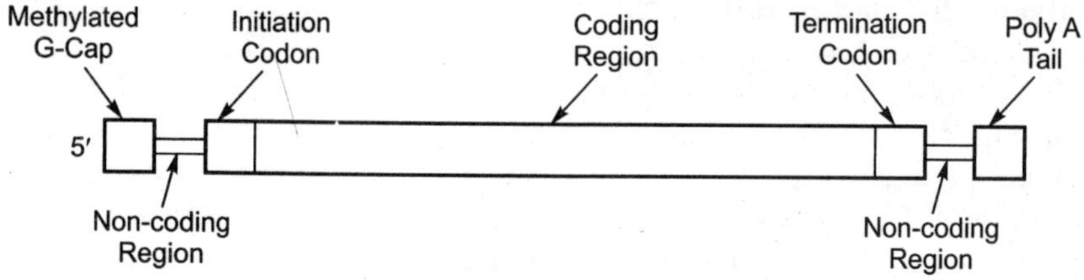

Fig. 4.6. Structure of mRNA strand.

Differences between DNA and RNA

DNA	*RNA*
1. DNA mainly occurs in the nucleus. A small quantity is found in mitochondria and plastids.	1. RNA mainly occurs in the cytoplasm. Very small quantity is found in the nucleus.
2. It is the genetic material.	2. It is not the genetic material except in certain viruses *e.g.*, Reovirus.
3. DNA is double stranded with the exception of some viruses (*e.g.*, $\phi \times 174$).	3. RNA is single stranded with the exception of some viruses (*e.g.*, double stranded in Reovirus).
4. It has regular helical coiling.	4. There is no regular coiling except in parts of tRNA.
5. DNA contains deoxyribose sugar.	5. RNA contains ribose sugar.
6. It contains nitrogen bases : adenine (A), guanine (G), thymine (T) and cytosine (C).	6. It contains nitrogen bases : adenine (A), guanine (G), uracil (U) and cytosine (C).
7. Hydrogen bonds are formed between complementary nitrogen bases of the opposite strands of DNA (A-T, C-G).	7. Base pairing through hydrogen bonds occur only in the coiled parts.
8. Purine and pyrimidine bases are in equal number.	8. There is no such proportionality between the number of purine and pyrimidine bases.
9. DNA contains over a million nucleotide.	9. Depending upon the type, RNA contains 70–12,000 nucleotides.
10. It can replicate to form new DNA.	10. It cannot replicate except in RNA-RNA viruses.
11. DNA transcribes genetic information to RNA.	11. RNA translates the transcribed message for the synthesis of polypeptides.
12. It is only of two types—nuclear and extra nuclear.	12. It is of three types–mRNA, rRNA and tRNA.
13. Its quantity is fixed in each cell of a species, except gametes.	13. Its quantity varies in different cells of a species.
14. DNA forms chromosomes or similar structure.	14. rRNA forms ribosomes.
15. It controls, heredity, evolution, metabolism, structure and differentiation.	15. It participates in protein synthesis and also help in the control of metabolism under instruction of DNA.

Functions of RNA

RNA play a multiple role in cells. These are :

- RNAs are involved in the expression of genetic code of DNA by forming specific proteins.
- RNA is the genetic material in some viruses.
- rRNA is a component of ribosomes, the site of protein synthesis.
- RNA primer is essential for DNA replication.
- Folding of prokaryotic DNA into a nucleoid needs the help of RNA.
- Some RNAs act as enzymes *e.g.*, ribozyme, ribonuclease.

REPLICATION OF DNA

Watson and Crick model not only depicts the structure of DNA but also shows that DNA molecule can replicate itself faithfully, and remains relatively stable from generation to generation and also explains how information is contained in DNA molecule. Watson and Crick were aware of this potential and wrote in the journal '*Nature*'. "It has not escaped our notice that the specific pairing we have postulated immediately suggested a possible copying mechanism for the genetic material."

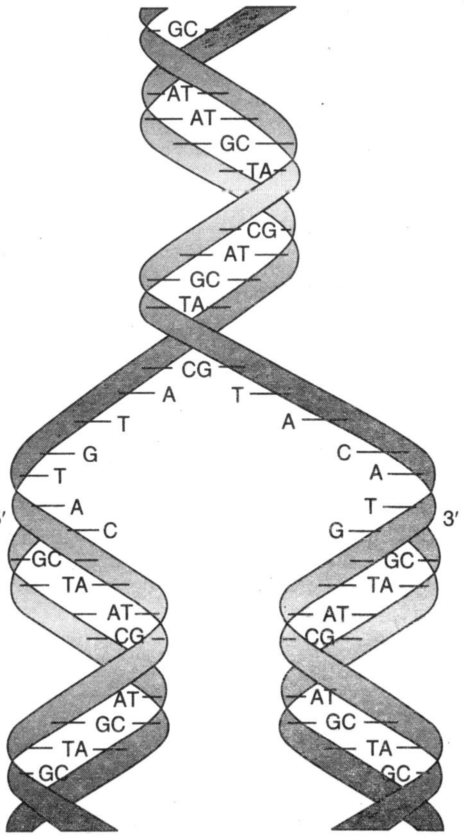

Fig. 4.7. Watson-Crick model for semi-conservative DNA replication.

Watson and Crick proposed that DNA replication is **semi-conservative**. It is a type of replication in which one strand of daughter duplex is derived from the parent, while the other strand is formed a new. The two strands of a DNA molecule separate and each strand serves as a template (model or guide) for the formation of a new but complementary strand. Thus, the new or daughter DNA molecules formed would be made of one old or parental strand and another newly formed complementary strand. This method of formation of new daughter DNA molecules is said to be semi-conservative (Fig. 4.7).

The Machinery and the Enzymes of DNA Replication

The process of replication is a complex, multi-step process which involves the participation of over a dozen of enzymes and other protein factors. These enzymes are highly efficient enzymes as they have to catalyze polymerization of a large number of nucleotides in a very short time. For instance, in *E. coli* that has only 4.6×10^6 bp, the process of replication completes within 38 minutes. That means the average rate of polymerization has to be approximately 2000 bp per second. In human beings where diploid content is 6.6×10^9 bp, the process of replication completes in 9 hours 15 minutes. Not only these enzymes to be fast, but they should catalyse the replication with high degree of accuracy. Any mistake during replication would result into mutations. The process of DNA replication involves following steps :

1. Origin of Replication. Replication of DNA begins at a specific point called **origin of replication** or **ori.** In most bacteria and viruses, there is only one origin of replication which moves bidirectionally. In eukaryotes, with giant DNA molecule, there can be several origins of replication, and these eventually merge with each other once replication is under process. Thus, a prokaryotic DNA acts as a single replicating unit or **replicon**, while a eukaryotic DNA molecule has several replicating segments or replicons.

2. Activation of Deoxyribonucleotides. The deoxyribonucleotide monophosphates-deAMP (deadenosine monophosphate), deGMP (deguanosine monophosphate), deCMP (decytidine monophosphate) and deTMP (dethymidine monophosphate) occur freely in the nucleoplasm. They serve as raw material in DNA synthesis. They are first phosphorylated and changed into active forms with the utilization of ATP. The reactions are catalysed by enzyme phosphorylase. The phosphorylated nucleotides are deATP (deadenosine triphosphate), deGTP (deguanosine triphosphate), deCTP (decytidine triphosphate), and deTTP (dethymidine triphosphate).

3. Unwinding of DNA Helix. The two strands of DNA molecule separate, before each serves as a template for the synthesis of a new strand. It is not easy to separate the interwind strands of a long DNA molecule. It is like separating two interwind ropes by pulling them apart with force. But the moment you stop applying force, the two strands will automatically interwine with each other. At the same time, if you cut partially separated strands, the tension is relieved and the two strands will not wind back together.

Enzymes **helicases** and **topoisomerases** almost work at the same principle, while unwinding the DNA molecule. Enzymes helicases (unwindases) unwind the DNA helix and separate the two strands, while the enzymes topoisomerases (DNA gyrases) can break and reseal one strand of DNA to facilitate uncoiling. The separated strands are stabilized by **helix destabilizing proteins.** The whole of DNA does not open at one stretch, but the separation of the two strands proceeds slowly from one end to another. It results in the formation of Y-shaped structure called **replication fork.**

4. Formation of RNA Primer. RNA primer is a short segment of RNA which is synthesized at the 5′ end of DNA before replication begins. The enzyme **primase** catalyses the polymerization of RNA building blocks (A, U, G, C) into primer. It is like putting a knot at the end of thread before you make a string of pearls. Formation of RNA primer constitutes the initiation phase of DNA synthesis because without the presence of RNA primer, DNA polymerase cannot initiate the synthesis of a new DNA strand. Later, the primer RNA, is removed and the gap is filled with deoxyribonucleotide to make DNA strand continuous.

5. Base Pairing. The two separated strands in the replication fork function as templates. The deoxyribonucleoside triphosphates come to lie opposite the appropriate nitrogen bases of both the exposed DNA templates according to base pairing rule *i.e.,* deATP opposite T, deGTP opposite C, deTTP opposite A and deCTP opposite G. The two extra phosphate present on the deoxyribonucleotides set free with the help of enzyme **pyrophosphatase** and energy is released. This energy is utilized in the formation of hydrogen bonds between the free nucleotides and nitrogen bases of the templates.

$$\text{Deoxyribonucleoside triphosphate} \xrightarrow{\text{Pyrophosphatase}} \text{Deoxyribonucleoside monophosphate} + 2\ \text{Pi} + \text{Energy.}$$

6. Elongation of New Strands. The adjacent deoxyribonucleoside monophosphates attached to two template DNA strands get linked to form replicated DNA strands. The process is catalysed by the enzyme **DNA polymerase III** and requires the presence of Mg^{2+}, ATP (or GTP) and TTP. DNA polymerase III along with other DNA polymerases (I and II) has the capability to elongate an existing DNA strand but can not initiate the synthesis. All the three DNA polymerases can polymerise nucleotides only in 5′ → 3′ direction bacause they add them at the 3′ end of the sugar molecule. The two strands of DNA run in antiparallel directions, therefore, the two templates provide different ends for replication. Replication over the two templates thus proceed in opposite directions. One strand is formed continuously because its 3′ end is open for elongation. It is called **leading strand**. On the other strand, short DNA segments are formed in the 5′ 3′ direction due to exposure of a small stretch of template at one time. The short segments of replicated DNA are called **okazaki fragments or okazaki segments** (named after Japanese scientist, **Okazaki**, who discovered them in 1968). Each of them has 3000–4000 nucleotides. An RNA primer is required every time a new okazaki fragment is to be formed. Such a process is also referred to as semi-discontinuous replication. The RNA primers are then removed and the gaps are filled by DNA synthesis. Both steps are performed by DNA polymerase I. The okazaki fragments are later joined up by the enzyme **DNA ligase (DNA synthetase)**. This strand is called **lagging strand** (*Fig. 4.8*). This mechanism, where leading strand is synthesized continuously and lagging strand is synthesized discontinuously, is called **semi-discontinuous replication.**

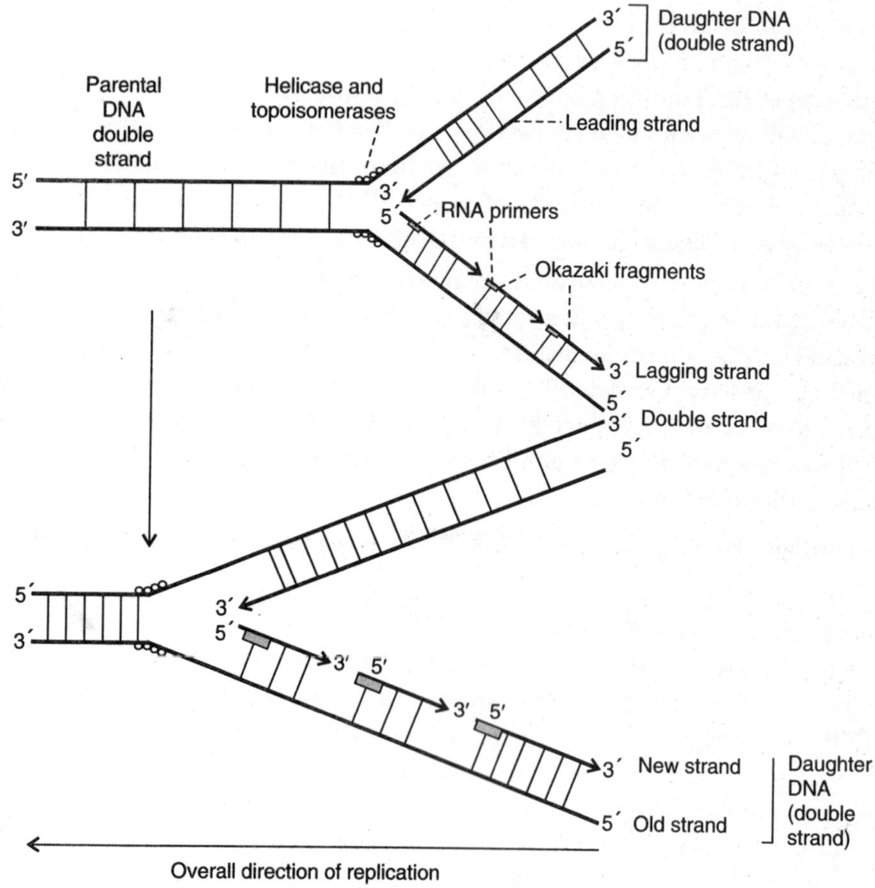

Fig. 4.8. Continuous and discontinuous synthesis of DNA. One new strand referred to as the leading strand is formed in the $5' \rightarrow 3'$ direction in a continuous fashion following the movement of fork. The other new strand is formed in small pieces, (referred to as okazaki fragments) and then joined together. The small pieces also grow is $5' \rightarrow 3'$ direction but the overall direction of the lagging strand is $3' \rightarrow 5'$.

PROOFREADING AND DNA-REPAIR

The specificity of base pairing ensures accurate replication. However, sometimes a wrong base may get in. The frequency is one in ten thousand. Bacterial DNA polymerase is able to sense the same. It go back, removes the wrong base and allows the addition of proper base before it proceeds forward.

There is also a repair mechanism for abnormal base pairs formed in the DNA due to mutation, escaping the proofreading mechanism. The damaged region of the DNA strand is corrected with the help of enzymes called **repair enzymes.** Enzyme **endonuclease** causes a break near the region of repair. **DNA polymerase I** and **III** remove the mismatched or abnormal nucleotides and synthesises a correct replacement by using the intact strand as template. The newly formed segment is sealed by **DNA ligase.** Proofreading and repair mechanisms ensures formation of identical DNA molecules during replication.

Unidirectional and Bidirectional DNA-Replication

John cairns (1963), from his autoradiography experiments on *E. coli,* described that DNA replication starts at a fixed point on the chromosome and proceeds in one direction, it is called **unidirectional replication**. However, recent studies revealed that in many prokaryotes and eukaryotes, DNA replication is **bidirectional**. In bidirectional replication, two replication forks are formed at the origin of replication and grow in opposite directions.

Differences between Leading Strand and Lagging Strand

Leading Strand	*Lagging Strand*
1. It is continuously formed replicated strand of DNA without any gap.	1. It is a replicated strand of DNA which is formed by the joining of short segments called okazaki segments.
2. The growth of leading strand is in 5′ – 3′ direction.	2. Each okazaki segment grows in 5′–3′ direction, but the growth of the lagging strand is in 3′–5′ direction.
3. It does not require DNA ligase for its growth.	3. DNA ligase is required for the formation of lagging strand from okazaki segments.

REVIEW EXERCISES

MULTIPLE CHOICE TYPE QUESTIONS

1. A nucleoside is formed of
(*a*) Pentose sugar, phosphate and nitrogen base
(*b*) Phosphate and nitrogen base
(*c*) Pentose sugar and phosphate
(*d*) Pentose sugar and nitrogen base.

2. Hydrolysis of nucleoside does not yield
(*a*) Pentose sugar (*b*) Phosphoric acid
(*c*) Pyrimidine (*d*) Purine.

3. Which one of the following is made up of double ring of atoms?
(*a*) Adenine (*b*) Thymine
(*c*) Cytosine (*d*) Uracil.

4. A short length of DNA molecule contains 120 adenine and 120 cytosine bases. The total number of nucleotides in this DNA segment is
(*a*) 60 (*b*) 120
(*c*) 240 (*d*) 480.

5. Adenine is a
(*a*) Nucleoside (*b*) Nucleotide
(*c*) Pyrimidine (*d*) Purine.

6. Guanine is 10% in a DNA sample, thymine would be
(*a*) 80% (*b*) 40%
(*c*) 10% (*d*) 90%.

7. The basic unit of nucleic acid is
(*a*) Pentose sugar (*b*) Phosphoric acid
(*c*) Nitrogenous base (*d*) Nucleotide.

8. Which one of the following pairs of nitrogenous bases of nucleic acids, is wrongly matched with the category mentioned against it?

(*a*) Adenine, Thymine–Purines
(*b*) Thymine, uracil–Pyrimidines
(*c*) Uracil, Cytosine–Pyrimidines
(*d*) Guanine, Adenine–Purines.

9. In a DNA molecule

(*a*) There are two strands which run antiparallel $5' \rightarrow 3'$ direction and other in $3' \rightarrow 5'$.
(*b*) The total amount of purine nucleotides and pyrimidine nucleotides is not always equal.
(*c*) There are two strands which run parallel in the $5' \rightarrow 3'$ direction.
(*d*) The proportion of adenine in relation to thymine varies with the organism.

10. The $3' - 5'$ phosphodiester linkages inside a polynucleotide chain serve to join.

(*a*) One nitrogen base with pentose sugar
(*b*) One DNA strand with the other DNA strand
(*c*) One nucleoside with another nucleoside
(*d*) One nucleotide with another nucleotide.

OBJECTIVE TYPE QUESTIONS

1. Fill in the blanks with suitable words:

(*i*) RNA has uracil instead of ....
(*ii*) A nucleotide molecule consists of three smaller molecules : pentose sugar, ... and ....
(*iii*) Double helix model of DNA, was proposed by .... and ....
(*iv*) One turn of DNA double helix measures .... and contains .... base pairs.
(*v*) The successive base pairs in DNA double helix are .... a part.

2. Match the items listed in Column I with appropriate items from Column II.

Column I	Column II
(*i*) F. Miescher	(*a*) Double helical model of DNA.
(*ii*) Altman	(*b*) Nuclein
(*iii*) Watson and Crick	(*c*) DNA as genetic material
(*iv*) Harshey and Chase	(*d*) Nucleic acid

TEST QUESTIONS

1. Name the monosaccharides found in RNA and DNA.
2. Name the two components of a nucleoside.
3. What is the difference between a nucleoside and nucleotide? Give two examples of each with their structure.
4. Give the salient features of Watson and Crick double helix model of DNA.
5. What is meant by semi-conservative nature of DNA replication?
6. What do you understand by proofreading and DNA repair? Explain.
7. Differentiate between leading and lagging strands.
8. List the main differences between DNA and RNA.
9. Describe the process of DNA replication.
10. Describe the three types of RNA along with their functions.

5 Protein Synthesis : Central Dogma

- Functions of Proteins
  - — Central Dogma
- Transcription
  - — Transcription Unit
  - — Transcription Unit and the Gene
  - — Types of RNA and the Process of Transcription
- Genetic Code
  - — Characteristics of Genetic Code
  - — Exceptions in Genetic Code
  - — DNA—The Master Copy
  - — Role of tRNA as an Adapter Molecule
- TRANSLATION (mRNA to Protein or Protein Synthesis)
  - — Machinary for Protein Synthesis
  - — Mechanism of Protein Synthesis
- Regulation of Gene Expression or Protein Synthesis
  - — Operon Concept
- Importance of Gene Regulation
- Review Exercises
- Test Questions

Besides replicating precisely, DNA is also able to transfer the genetic information stored in it for the synthesis of polypeptides or proteins. The proteins are the most important components of the cell. They are the building materials of the protoplasm, and perform a variety of functions in the cell.

Proteins are polypeptides. They are linear chains of amino acids linked by peptide bonds. A protein may be a heteropolymer or homopolymer. A homopolymer has only one type of monomer repeating '*n*' number of times. The linear polymers of amino acids are called polypeptides. A protein may contain one or more polypeptides. In a polypeptide, many amino acids remain serially linked by peptide bond (—CO—NH—) formed between amino and carboxylic groups of successive amino acids.

$$H_2N{-}\overset{\overset{\displaystyle R_1}{|}}{\underset{\underset{\displaystyle H}{|}}{C}}{-}CO\boxed{OH + H}NH{-}\overset{\overset{\displaystyle R_2}{|}}{\underset{\underset{\displaystyle H}{|}}{C}}{-}COOH \longrightarrow H_2N{-}\overset{\overset{\displaystyle R_1}{|}}{\underset{\underset{\displaystyle H}{|}}{C}}{-}\underbrace{CO{-}NH}_{\text{Peptide bond}}{-}\overset{\overset{\displaystyle R_2}{|}}{\underset{\underset{\displaystyle H}{|}}{C}}{-}COOH + H_2O$$

The various permutations and combinations of the 20 amino acids allow the synthesis of a large number of proteins. For instance, in a polypeptide of only 100 amino acid residues, there can be 20^{100} possible arrangements or types polypeptides. This accounts for thousands of specific proteins found in each of the living species.

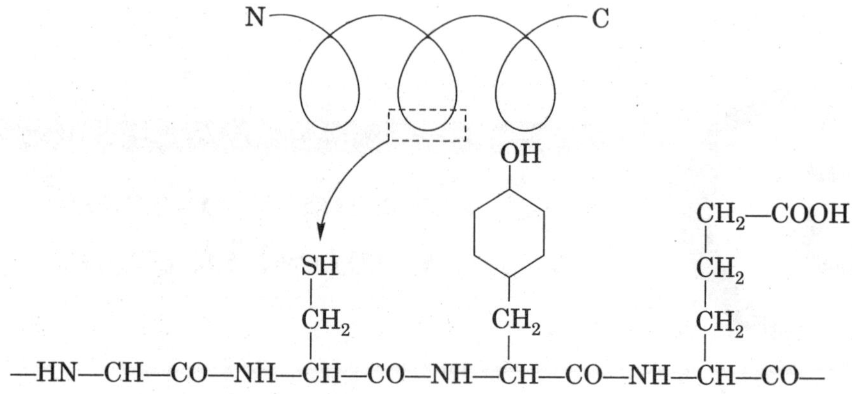

Amino acid Residues Serine — Cysteine — Tyrosine — Glutamic acid
(Ser) (Cys) (Tyr) (Glu)

Fig. 5.1. A portion of a polypeptide. N and C are the two termini of a polypeptide.

FUNCTIONS OF PROTEINS

Proteins perform a number of functions in the organisms. The main functions of proteins are as follows:

- **Structural Proteins.** Many proteins serve as building material of cells and tissues. They take part in the formation of colloidal complex of protoplast, cell membranes organelles, extracellular matrices and fibres. Some proteins form supporting structures, *e.g.*, elastin of ligaments, collagen of tendons, cartilages, bone and connective tissue.
- **Protective Structures.** Fibrous protein keratin is the major constituent of external protective structure of animals like hair, feathers, horny layer of skin, nails, claws, hoofs, etc. Cobweb of spider and protective covering in the cocoon stage of silk moth is made of white insoluble protein fibroin.
- **Compatibility Proteins.** The walls of pollen grain contain proteins for compatibility, incompatibility reaction with the stigma during pollination.
- **Enzymes.** Many proteins function as enzymes to catalyse biochemical reactions that occur in the living world. Enzymes play a key role in the metabolism.
- **Carrier Proteins (Transport Proteins).** Some proteins act as carriers which bind and transport specific molecules across a membrane or in a body fluid. Haemoglobin of RBCs transports oxygen in the body. Myoglobin of muscles store oxygen. α-globulin of blood carries thyroxine and bilirubin, β-globulin transport vitamins A, D and K, cholesterol and ions in the blood. Serum albumin carries fatty acids and lipids in the blood. In plants P-protein helps in the transport of organic compounds through phloem.
- **Receptor Proteins.** A number of proteins present on the exernal surface of cell membrane act as receptor molecules. These bind with specific informational molecules like hormones reaching the cell and mediate in their cellular effects.
- **Hormones.** Some hormones are proteinaceous, *e.g.*, insulin (sugar metabolism regulating hormone) parathyroid (calcium and phosphate transport regulating hormones). Hormones play a key role in the regulation of metabolism.

- **Contractile Proteins.** Myosin and actin make the muscle fibres contractile to bring about movements and locomotion. The contractile system is basically made up of protein actin but association with myosin is essential for contraction.
- **Defence.** Some proteins act as antibodies (immunoglobulins) that participate in the defence mechanism of the body.
- **Microtubules.** They are formed of protofilaments built up of two proteins, α and β tabulin. Microtubules are unbranched, hollow. Submicroscopic tubules which form the structural material of cilia, flagella, basal bodies, centrioles and spindle apparatus.
- **P-protein.** It is a special vibratile protein present in sieve tube elements, which is supposed to actively participate in the transport of nutrients.
- **Storage Proteins.** These occur in milk, eggs and seeds to nourish the young ones. They include casein of milk, **albumin** (ovalbumin) of egg white and **glutelin** in cereals. Iron storing protein commonly found in animal tissue is **ferritin.** The proteins having all the essential amino acids are called **first class proteins.**
- **Protein Buffers.** Proteins also help in maintaining a balance of acidity and alkalanity by combining with excess acids and bases. They, thus, act as 'biological buffers'.
- **Visual Pigments.** Rhodopsin and iodopsin are protein pigments. They are present in rods and cones of retina respectively. They take part in perception of image.
- **Toxins.** Many toxins of microbes, plants and animals are proteins. Bacterial toxins, ricin of castor oil seeds and snake venom are proteins.
- **Blood Clotting Proteins.** The proteins fibrinogen and thrombin help in blood clotting to check bleeding from injuries.
- **Mucoproteins.** The mucoproteins are present in mucus produced by salivary glands and mucus glands of an alimentary canal. The mucus protects the lining of alimentary canal from friction and digestive juices.
- **Sweetest Substance.** Monellin a protein derived from an African berry is 2000 times sweeter than sucrose. It is non-toxic and non-caloric, hence may prove suitable for diabetic patients.
- **Repressor.** Most of the repressors that regulate gene action are protein in nature.
- **Other Proteins.** Enkephalins are small peptides, which bind to specific receptor sites in the brain and influence the perception of pain and pressure. Human memory is believed to be stored in specific proteins called **memory proteins.**

Central Dogma

The proposition of 'double helix' structure of DNA and its simplicity in explaining the genetic implication became revolutionary. Very soon **Francis Crick** (1958), proposed the central dogma in molecular biology, which state that *'the genetic information flows from DNA to RNA and then to protein.* The concept of central dogma emphasized that three processes are involved in the flow of information from DNA. These are: (*i*) **Replication** (*i.e.,* exact copying or duplication of DNA), (*ii*) **Transcription** (*i.e.,* transfer of genetic information from DNA to mRNA), and (*iii*) **Translation** (*i.e.,* conversion of genetic information of mRNA in the form of protein).

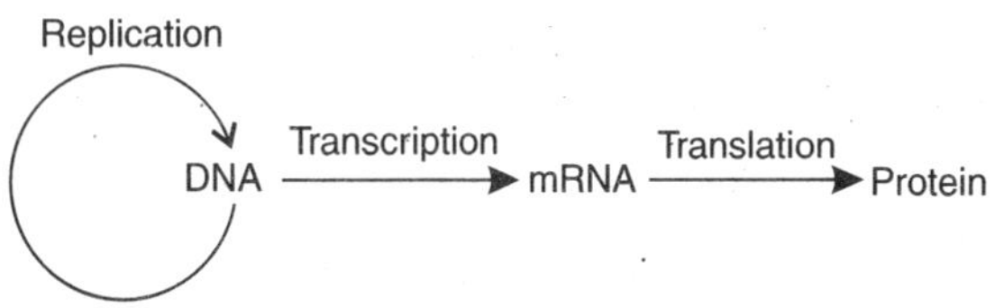

An exception to one way flow of information was reported by **H. Temin** and **D. Beltimore** (1970) in some viruses. They discovered that many tumor viruses contain RNA as genetic material and replicate by first synthesizing a complementary DNA. This process is called **reverse transcription**. It is carried out by an RNA-dependent DNA polymerse called *'reverse transcriptase'*. These viruses are called **retroviruses** and include Human Immunodeficiency Virus (HIV) that causes AIDS. This discovery in molecular biology introduced the concept of **'central dogma reverse'.** It states that the flow of genetic information is not necessarily from DNA to RNA but may be from RNA to DNA also.

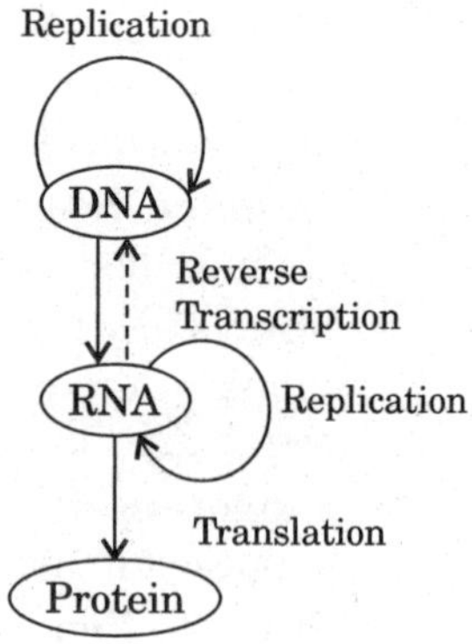

Mechanism of protein synthesis

TRANSCRIPTION

The process of copying genetic information from one strand of the DNA into RNA is called transcription. The process of transcription is also governed by the principle of complementarity. However, here, adanosine forms base pair with uracil instead of thymine. During transcription only a segment of DNA and only one of its strand is copied into RNA. It is because, if both strands are copied, they would code for RNA molecule with different sequences (because complementarity does not mean identical). Consequently, the sequences of amino acids in proteins synthesized by the two strands of a DNA segment would be different. This would complicate the genetic information transfer machinery. Secondly, if two RNA molecules produced simultaneously, they would be complementary to each other and would form a double stranded RNA. This would prevent RNA from being translated into protein, and the process of transcription would be of no use.

Transcription Unit

A transcription unit consists of three regions in DNA. There are— 1. A promoter, 2. The structural gene and 3. A terminator.

The two strands of the DNA in the structural gene forms a transcription unit. As the two strands of DNA have opposite polarity and the DNA-dependent RNA polymerase also catalyze the polymerization in only one direction, *i.e.*, $5' \rightarrow 3'$. Therefore, the DNA strand that has the polarity $3' \rightarrow 5'$ acts as a template. It is called **'template strand'.** The other strand which has the polarity $5' \rightarrow 3'$ and the sequence same as RNA (except thymine at the place of uracil), is displayed during transcription. Strangely, this strand (which does not code for anything) is termed as **'coding strand'.** All the reference point, while defining a transcription unit is made with coding strand. This can be explained by taking a hypothetical sequence from a transcription unit.

3′-ATGCATGCATGCATGCATGCATGC-5′ Template Strand

5′-TACGTACGTACGTACGTACGTACG-3′ Coding Strand

The sequence of RNA transcribed from above DNA will be as under

5′-UACGUACGUACGUACGUACGUACG-3′

The promoter and the terminator regions of a transcription unit are located at either ends of the structural genes. The promoter region is present toward 5′ end (upstream), whereas the terminator region is located toward 3′ end (down stream) of the structural gene (the reference of 5′ end and 3′ end is made with respect to the polarity of coding strand). The terminator region defines the end of the process of transcription. The presence of a promoter in a transcription unit defines the template and coding strands (Fig. 5.2). The switching of promoter with terminator could reverse the definition of coding and template strands. A number of regulatory sequences may be present further upstream or downstream to the promoter.

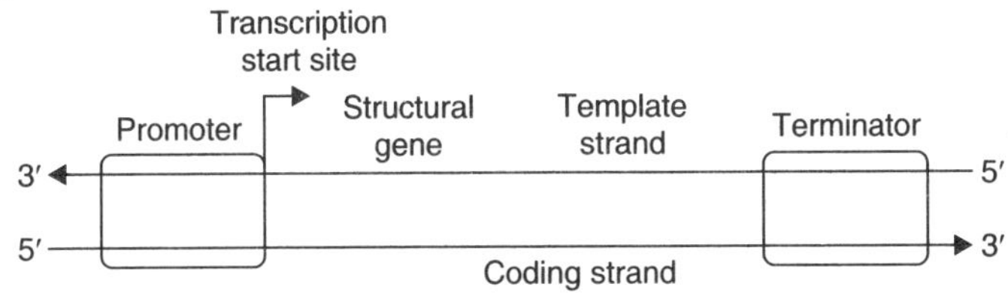

Fig. 5.2. Schematic structure of a transcription unit.

Transcription Unit and the Gene

DNA not only contains information for the synthesis of polypeptides through mRNA but the synthesis of other RNAs (tRNA or rRNA) as well. Therefore, the term gene, nowadays being replaced by **cistron.**

Cistron can be defined as a segment of DNA capable of directing the synthesis of a polypeptide chain or a RNA molecule.

While defining a cistron as segment of DNA coding for a polypeptide, the structural gene in a transcription unit could be said as **monocistronic** (mostly in eukaryotes) or **polycistronic** (mostly in prokaryotes such as bacteria). In eukaryotes, the monocistronic structural gene has two types of interspersed coding and non-coding segments called **exons** and **introns** respectively. Such a gene is called **split gene (Roberts** and **Sharp,** 1993). Exons are essential regions of a gene which on transcription form part of *m*RNA that code for different regions of protein. Introns are non-essential regions of the gene which are also called **junk DNA** or **spacer DNA** or **intervening sequence (IVS).**

During the synthesis of mRNA, the junk DNA or introns also get transcribed. Thus a long strand of mRNA is synthesised initially. It contains both unwanted base sequences (transcribed from introns) alternated with useful base seqences (transcribed from exons). Such a strand of mRNA is called **heterogeneous nuclear RNA** (hnRNA) or **pre-messenger RNA.** The unwanted base sequences are now removed by enzymes from hnRNA. The remaining useful base sequences coding for amino acids joint together. This processing of mR*NA* is known as **splicing.** Thus, the processed mRNA and the polypeptide it codes for are collinear even in the eukaryotes, although the genes are split. The processed *m*RNA passes out of the nucleus and joins the ribosomes. Here, it is expressed through the synthesis of a polypeptide.

The introns or intervening sequences do not appear in mature or processed RNA. The split gene arrangement further complicates the definition of a gene in terms of a DNA segment. The inheritance of a character is also affected by promoter and regulatory sequences of a structural gene. Hence, sometime the regulatory sequences are loosely defined as **regulatory genes,** even though these sequences do not code of any RNA or protein.

Types of RNA and the Process of Transcription

There are three types of RNAs : messenger RNA (mRNA), transfer RNA (tRNA) and ribosomal RNA (rRNA). All the three types of RNAs are required to synthesize a protein in a cell. During the process of translation, mRNA provides the template, tRNA brings amino acids at the site of polypeptide formation (*i.e.*, at the ribosome) and reads the genetic code, and rRNAs play structural and catalytic role.

1. Ribosomal RNA (rRNA). It is the constituent of ribosomes. Ribosomes are the ribonucleoprotein particles that provide the sites for protein synthesis. On the basis of their sedimentation coefficients, prokaryotic and eukaryotic ribosomes are characterised as 70 S and 80 S, respectively. Both types of ribosomes consist of a small and a large subunit (Figs. 5.3 and 5.4). In prokaryotes, the large or 50 S subunit consists of 23 S rRNA and a 5 S rRNA + 32 different proteins and the 30 S small subunit is made up of 16 S rRNA and 21 different proteins. On the other hand, the eukaryotic large subunit (60 S), is composed of 28 S rRNA, 5 S RNA and 5.8 S rRNA and 50 proteins and the 40 S small subunit consists of 18 S rRNA and 33 different proteins.

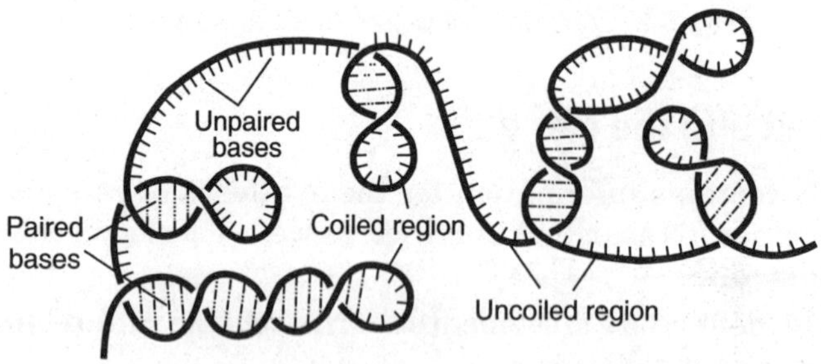

Fig. 5.3. Structure of rRNA (schematic).

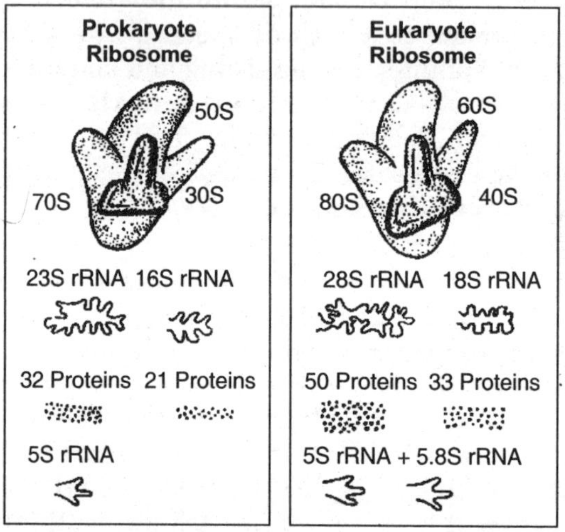

Fig. 5.4. Generalized structure of ribosome in prokaryotes and eukaryotes.

2. Transfer RNA (tRNA). It is also called double RNA or sRNA. Transfer RNA constitutes about 15% of total RNA. tRNA molecules occur in the cytoplasm and are smallest of the three types. They carry amino acids to the ribosomes during translation and hence act as **adapter molecules.** tRNA contains a number of modified bases. tRNA is single stranded but folded on to itself (Fig. 5.5) giving rise to clover leaf like form in two dimension (**Holley**, 1964) and an inverted L-shaped form in three dimension (**Kim,** 1973).

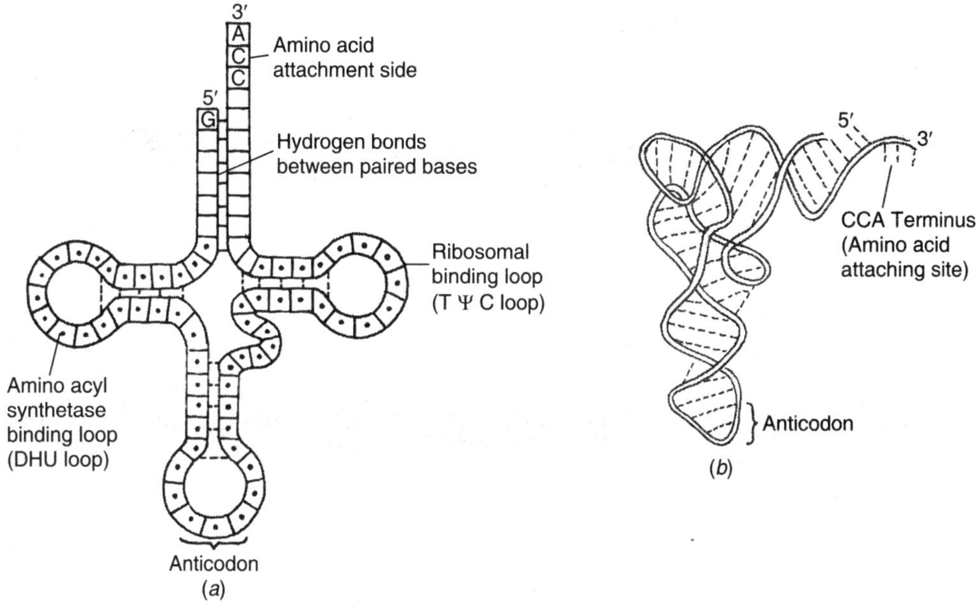

Fig. 5.5. Structure of transfer RNA;
(*a*) The transfer RNA is a folded molecule and there are four double stranded regions formed by hydrogen bonding between bases giving rise to three loops. The amino acid attaches to the 3′ end of the transfer RNA molecule. The anticodon interacts with the codon on the messenger RNA and (*b*) In three dimension, transfer RNA looks like an inverted L-shaped molecule.

Transfer, RNA has a site for binding amino acid. It lies at 3′ end and has unpaired CCA sequence with —OH at the tip. The carboxylic (—COOH) group of amino acid joins the —OH. tRNA has three loops—anticodon loop, ribosomal binding loop and dihydroxy uridine loop.

(*i*) ***Anticodon Loop.*** It lies opposite to amino acid or AA binding site and has an anticodon made up of three nitrogen bases for recognising and attaching to the codon of mRNA. The anticodon of tRNA reads its appropriate codon and temporarily joins it by hydrogen bonds during protein synthesis.

(*ii*) ***Ribosomal Binding or T ψ C Loop.*** This loop is meant for attaching to the ribosome.

(*iii*) ***Dihydroxy Uridine or DHU Loop.*** It is meant for binding to the enzyme-amino-acyl-synthetase.

An extra or variable arm may be present near the T ψ C loop. Transfer RNA is meant for transferring amino acids to ribosomes for synthesis of polypeptides. There are different tRNA for different amino acids. Some amino acids can, however, be picked up by 2 to 6 tRNAs. Thus, there are over 100 types of tRNA in a cell. tRNAs are synthesised as usual on DNA templates. However, their 3′ end having CCA sequence is believed to be added after transcription.

3. Messenger RNA (mRNA). Messenger RNA is the longest of all RNAs. It constitutes only 5% of the total RNA content of the cell. mRNA carries genetic information from DNA for the formation of particular polypeptide. The information is present in the base sequence of its nucleotides. It is called **genetic code.** The sequence of nitrogen bases on mRNA form codons to specify amino acids in polypeptide chain. The 5′ terminus of mRNA has a methylated region, which acts as cap for attachment with ribosome. An initiator or start codon (AUG) follows the cap either immediately or after a small non-coding region. Then, there is coding region, it is followed by terminator codon (UAA, UAG or UGA), a small non-coding region and poly A area at the 3′ terminus. In eukaryotes, the mRNA usually specify a single polypeptide. Such mRNA is called **monocistronic.** In prokaryotes, single mRNA may specify more than one polypeptide. Such mRNA is termed as **polycistronic.**

In prokaryotes such as bacteria, there is single DNA dependent RNA polymerase that catalyzes transcription of all the three types of RNAs.

The process of transcription involves following steps (Fig. 5.6):

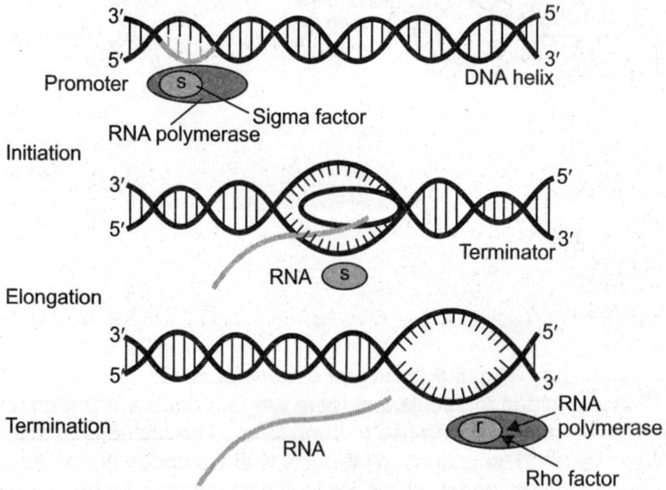

Fig. 5.6. Process of transcription in Bacteria.

(*i*) ***Initiation.*** On specific signal, the RNA polymerase binds to promoter and initiates transcription. The promoter region in the cistron is recognised by sigma (σ) subunit of the enzyme RNA polymerase in prokaryotes and by many transcription factors in eukaryotes.

(*ii*) ***Elongation.*** The RNA polymerase uses ribonucleoside triphosphates as substrate and polymerizes in a template dependent manner following the rule of complementarity. RNA polymerase moves along the transcription unit and causes unwinding and opening of the helix and continues the elongation of RNA. The sigma (σ) factor of the RNA polymerase separates on the initiation of RNA chain. The remaining part of the RNA polymerase, referred as core enzyme moves along the DNA template causing elongation of RNA chain at the rate of some 30 nucleotides per second. Elongation takes place in 5′–3′ direction.

(*iii*) ***Termination.*** Once the RNA polymerase reaches to the terminator, the nascent RNA fall off and also the enzyme, a terminator factor—Rho (ρ) factor is required for the termination of RNA chain. Unlike the situation in prokaryotes, transcription in eukaryotes occurs within the nucleus and mRNA moves out of the nucleus into the cytoplasm for translation. In eukaryotes, there are two additional complexities:

1. There are at least three RNA polymerases in the nucleus (in addition to the RNA polymerase found in the organelles). There are RNA polymerase I, II and III. The RNA polymerase I transcribes rRNAs (28S, 18S and 5.8S), whereas RNA polymerase III is responsible for transcription of tRNA, RNA, 5*s* rRNA and *sn* RNAs (small nuclear RNAs). The RNA polymerase II transcribes precursor of mRNA (pre-mRNA) or hnRNA (heterogenous nuclear RNA). Thus, there is a division of labour in the functioning of the three types of RNA polymerase.

2. The second complexity is that the primary transcript contains two types of segments, the non-coding **introns** or **intervening sequences** and the coding **exons.** The primary eukaryotic mRNA transcript is much longer and is non-functional. Hence, it is subjected to a process called **'splicing'**, where the introns are removed and exons are joined in a definite order. Hn RNA (primary mRNA transcript) undergo two additional processing called as **'capping'** and **'tailing'.** In capping an unusual nucleotide, **methyl guanosine triphosphate** ($^{m}G_{ppp}$) is added to the 5′ end of hn RNA. In 'tailing' adenylate residues (200–300) are added (polyadenylation) at 3′-end to hn RNA in a template independent manner (*i.e.,* without a template). The fully processed hnRNA is now called as mRNA, that is transported out of the nucleus for translation (Fig. 5.7).

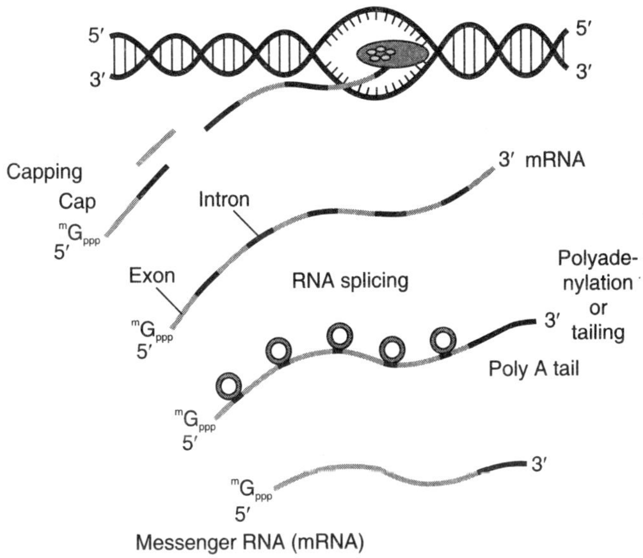

Fig. 5.7. Process of transcription of Eukaryotes.

Recently the significance of complexity in eukaryotic gene is being studied. It is believed that the split gene arrangement probably represent an ancient feature of the genome. The presence of introns is reminiscent of antiquity, and the process of splicing represents the dominance of 'RNA world'. As such the RNA and RNA dependent processes have assumed more importance in the living system.

GENETIC CODE

The process of translation requires transfer of genetic information from a polymer of nucleotides to a polymer of amino acids. But, neither there exists any complementarity between nucleotides and amino acids, nor could be drawn theoretically. However, there were enough evidences which support the view that change in nucleic acids (genetic material) were responsible for change in amino acids in proteins. This led to the proposition of a genetic code that could direct the sequence of amino acids during synthesis of proteins. *'The relationship between the sequence of amino acids in a polypeptide and nucleotide sequence of DNA or mRNA is called* **genetic code'**. The process by which the information coded in mRNA is decoded into a polypeptide is one of the exciting discoveries of biology. It is often referred to as deciphering the genetic code.

It was **George Gamow** (1954), a physicist, who argued that since there are only 4 bases and if they have to code for 20 amino acids, the code should constitute a combination of bases. In order to code for all the 20 amino acids, he suggested that the code should be made up of three nucleotides (**triplet code**). The permutation and combination of three nucleotides 4^3 ($4 \times 4 \times 4$) would generate 64 codons. Proving that codon was triplet (*i.e.,* three nucleotides) was quite challanging task. But the chemical method developed by **Har Govind Khurana** for synthesizing RNA molecules with defined combinations of bases (homopolymers and copolymers), and **Marshall Nirenberg's** cell free system for protein synthesis finally helped the genetic code to be deciphered. The discovery of enzyme-polynucleotide phosphosylase, by **Severe Ochoa** was also helpful to polymerise RNA with defined sequences in template independent manner (*i.e.,* enzymatic synthesis of RNA without a template). Finally a checker board for genetic code (Fig. 5.8) was prepared which is as follows.

FIRST BASE	SECOND BASE: U	C	A	G	THIRD BASE
U	UUU Phe	UCU Ser	UAU Tyr	UGU Cys	U
U	UUC Phe	UCC Ser	UAC Tyr	UGC Cys	C
U	UUA Leu	UCA Ser	UAA Stop (ochre)	UGA Stop (Opal)	A
U	UUG Leu	UCG Ser	UAG Stop (amber)	UGG Trp	G
C	CUU Leu	CCU Pro	CAU His	CGU Arg	U
C	CUC Leu	CCC Pro	CAC His	CGC Arg	C
C	CUA Leu	CCA Pro	CAA Gln	CGA Arg	A
C	CUG Leu	CCG Pro	CAG Gln	CGG Arg	G
A	AUU Ile	ACU Thr	AAU Asn	AGU Ser	U
A	AUC Ile	ACC Thr	AAC Asn	AGC Ser	C
A	AUA Ile	ACA Thr	AAA Lys	AGA Arg	A
A	AUG Met or Start	ACG Thr	AAG Lys	AGG Arg	G
G	GUU Val	GCU Ala	GAU Asp	GGU Gly	U
G	GUC Val	GCC Ala	GAC Asp	GGC Gly	C
G	GUA Val	GCA Ala	GAA Glu	GGA Gly	A
G	GUG Val	GCG Ala	GAG Glu	GGG Gly	G

Fig. 5.8. Genetic code. The first, second and third bases as read from 5′ to 3′ direction constitute the triplet code in RNA. The codon AUG specifies methionine and is usually the starting point for protein synthesis. The word 'stop' indicates codons serving as signals to terminate protein synthesis. For each amino acid more than one codon have been identified. It would be clear from the Figure that while the first and second bases remain the same for a particular amino acid, the third base can be different.

Characteristics of Genetic Code

1. Triplet code. The genetic code is a triplet code. Three adjacent nitrogen bases constitute a codon, which specifies the placement of one amino acid in a polypeptide.

2. Non-overlapping. The adjacent codons do not overlap. Each single base is a part of only one codon.

3. Commaless. The genetic code is continuous and does not possess pauses after triplets. Reading of the code begins at a fixed point and continues three nucleotides at a time, without a pause till the terminator codon.

4. Universality. The genetic code is universal, *i.e.,* a given codon in the DNA and mRNA specifies the same amino acid in the protein synthesizing systems of all organisms.

5. Non-sense or Terminator codons. Three of the 64 codons, namely UAA (ochre), UAG (amber) and UGA (opal) do not specify any amino acid but signal the end of the message. They are called **non-sense** or **terminator codons.** Either of these stops synthesis of the polypeptide chain.

6. Initiation or Start codons. The codons AUG and GUG code for methionine and valine respectively, also act as start signal (dual functions). They initiate the synthesis of polypeptide chain, hence are called **initiate** or **start codons.**

7. Degeneracy. The genetic codons are degenerate codons *i.e.,* they lack specificity and one amino acid often has more than one code triplet, only methionine and tryptophan have single triplet codons. All other amino acid are specified by 2–6 codons. Such codons are called degenerate codons. In degenerate codons the first two nitrogen bases are similar, while the third one is different. The third nitrogen base has no effect on coding, and it permit interaction with the anticodon on the corresponding tRNA. This position on the codon is known as the **wobble position.**

8. Colinearity. DNA is a linear polynucleotide chain and a protein is a linear polypeptide chain. The sequence of amino acids in a polypeptide chain corresponds to the sequence of nucleotide bases in the DNA (gene) that code for it. Change in specific codon in DNA produce a change of amino acid in the corresponding position in the polypeptide. The gene and the polypeptide it codes for are said to be **colinear.**

9. Gene polypeptide parity. A gene or a portion of DNA called **cistron** specifies the synthesis of a particular polypeptide. It means that the genetic system should have as many cistrons (genes) as the types of polypeptides found in the organism.

–If the sequence of nucleotide in an mRNA is known, the sequence of amino acids coded by it can be predicted by taking the help of checker board of genetic code.

Let us find out the sequence of amino acids for the transcript of an mRNA given below.

AUG UUU UUC UUC UUU UUU UUC

The sequence of amino acids coded by the above transcript will be—Met-Phe-Phe-Phe-Phe-Phe-Phe

Similarly, the sequence of nucleotides in an mRNA transcript can be predicted, if the sequence of amino acids in a polypeptide chain is known.

Exceptions in Genetic Code

1. Overlapping genes. Certain bacteria and animal viruses have a few overlapping genes. In $\phi \times 174$ virus, three genes E, B and K overlap other genes. Nucleotide sequence at the beginning of E gene is contained within gene D. Likewise gene K overlaps with gene A and C. A similar condition is found in SV-40.

2. Mitochondrial genes. The codons AGG and AGA which code for arginine, act as stop signals in human mitochondria. UGA, a termination codon codes for tryptophan while the codon AUA (for isoleucine) denotes methionine in human mitochondria.

3. Different codons. In *Paramoecium* and other ciliates termination colons UAA and UGA code for glutamine.

DNA—The Master Copy

DNA is called the master copy of genetic information, because it is not directly involved in the formation of polypeptide. DNA is carefully preserved in the nucleus and is used in the formation of the **working copies** in the form of mRNA molecules. The latter pass out to the ribosome in the cytoplasm to guide the synthesis of polypeptide.

Role of tRNA as an Adapter Molecule

Francis Crick (1961), who contributed significantly in deciphering the genetic code, suggested that there has to be a mechanism to read the code and also to link it to the amino acids, because amino acids have no structural uniqueness to read the code uniquely. He postulated an adapter molecule that would on one hand read the code and on other hand would bind to specific amino acids. It was found to be tRNA, which acts as an **adapter molecule.** tRNA has an 'anticodon loop' that has bases complementary to the code, and it also has an 'amino acid accepter end' to which it binds to amino acids. tRNAs are specific for each amino acids (Fig. 5.9). For initiation, there is another specific tRNA that is referred to as initiator tRNA. There are no tRNAs for stop codons.

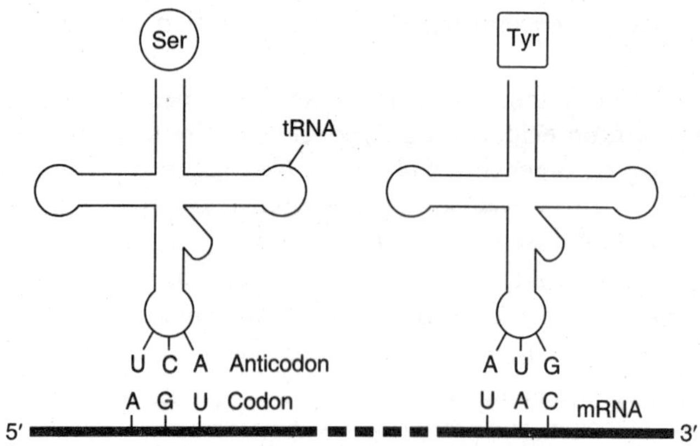

Fig. 5.9. tRNA—the adapter molecule.

TRANSLATION (mRNA to Protein or Protein Synthesis)

Translation refers to the process of polymerization of amino acids to form a polypeptide. The order and sequence of amino acids are defined by the sequence of mRNA. The amino acids are jointed by a bond known as **peptide bond.**

Machinary for Protein Synthesis

The synthesis of protein occurs on ribosomes and involve mRNA, tRNAs, amino-acyl tRNA synthetases and amino acids.

1. Ribosomes. Ribosomes are ribonucleo-protein particles that provide the sites for protein synthesis. Each ribosome consists of two unequal subunits, small and large. The smaller subunit fits over the larger one like a cap but leaves a tunnel for mRNA. The larger subunit has a groove for pushing out the newly formed polypeptide and protecting the same from cellular enzymes (Fig. 5.10). The two subunits come together only at the time of protein formation. Mg^{2+} ion is essential for the association of the two subunits of ribosome. During active protein synthesis a number of ribosomes are held on a strand of mRNA. Such a structure is called polyribosome or polysome. Polyribosome helps to produce a number of copies of the same polypeptide.

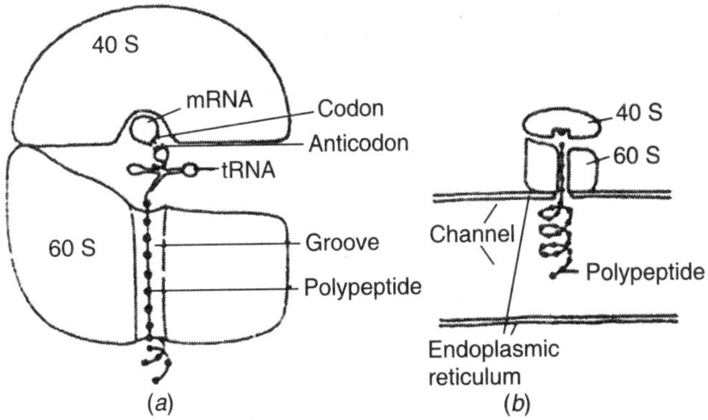

Fig. 5.10. Ribosomes as protein factory; (*a*) relationship between the various components and (*b*) Synthesis of polypeptide on ribosome connected with endoplasmic reticulum.

There are two reactive sites, P or D-site (peptidyl transfer or donor site) and A-site (amino-acyl or acceptor site). P-site is jointly contributed by the two ribosomal subunits. A-site is situated on the large subunit of ribosome. It faces the tunnel between the two subunits. Enzyme peptidyl transferase is a ribozyme and is a component of larger subunit. The smaller subunit of ribosome also has a site for recognising mRNA and binding area for initiation factors.

2. mRNA. It brings coded information from DNA for the synthesis of a specific polypeptide. The information is present in the base sequence of its nucleotides. A sequence of three nucleotides constitute a codon. The codons of mRNA are not recognised by amino acids but by the anticodons of their adapter molecules, the tRNAs.

3. tRNAs. They carry amino acids from the cellular pool to the site of translation. A tRNA picks up a specific amino acid at its CCA or 3' end. This process is called ***charging.*** The charged tRNA take the amino acid to mRNA over a specific codon corresponding to its anticodon. Each tRNA has an area for coming in contact with ribosome and the enzyme amino-acyl tRNA synthetase. A tRNA can pick up only a specific amino acid though an amino acid can be specified by 2–6 tRNAs.

4. Amino-Acyl tRNA Synthetase. (aa-activating enzyme). It helps in binding of amino acid to its specific tRNA.

5. Amino Acids. There are some 20 amino acids and amides in the cellular pool, which constitute the building blocks or monomers of protein. These amino acids and amides form hundreds of different types of proteins in a cell.

Mechanism of Protein Synthesis

The synthesis of protein is a complex process and involves following steps:

1. Activation of Amino Acids (Fig. 5.11). There are some 20 different kinds of amino acids which constitute building blocks or monomers of proteins. Though all types of proteins are formed from the same amino acids but the number and the arrangement of amino acids in the proteins provide specificity to the latter.

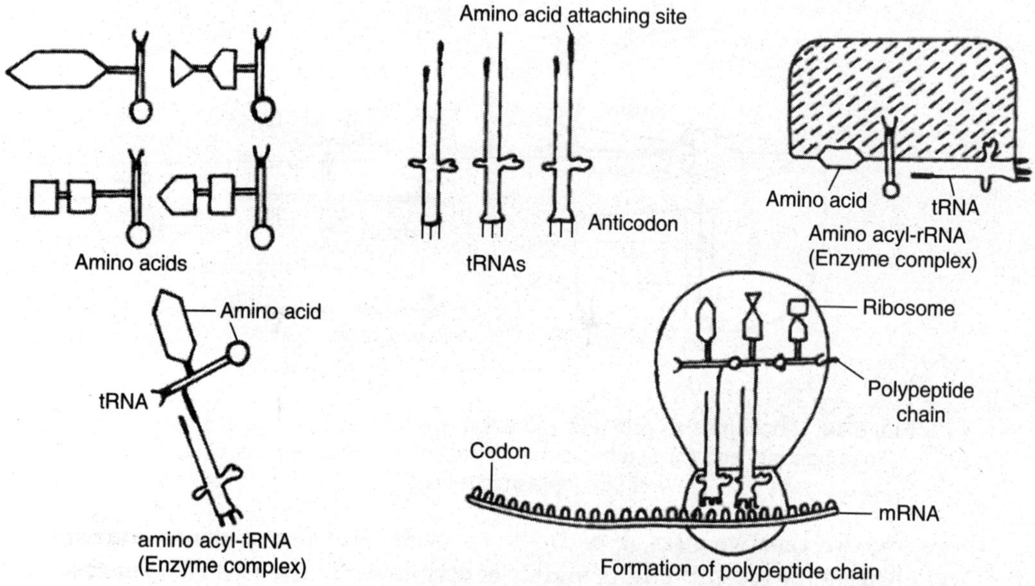

Fig. 5.11. Synthesis of polypeptide from amino acids involving tRNA, mRNA, amino acyl synthetase and ribosome.

Amino acids react with ATP to form **amino acid-AMP-complex** and inorganic phosphates. The reaction is catalysed by a specific amino acid activating enzyme called *amino acyl-tRNA synthetase* in presence of Mg^{2+}. The amino acid AMP complex remains temporarily associated with the enzyme. The amino acid AMP enzyme complex is called an **activated amino acid.** It reacts with tRNA specific for the amino acid to form amino acyl tRNA complex. The enzyme and AMP are released. The tRNA complexed with amino acid is called charged tRNA.

$$\underset{\text{(Amino acid)}}{\text{AA}} + \text{ATP} + \underset{\text{(Amino acyl tRNA synthetase)}}{\text{Enzyme}} \xrightarrow{Mg^{2+}} \text{AA-AMP-Enzyme} + \underset{\text{(Pyrophosphate)}}{\text{PPi}}$$

$$\text{AA-AMP-Enzyme} + \text{tRNA} \longrightarrow \underset{\text{(Aminoacyl tRNA complex)}}{\text{AA-tRNA}} + \text{AMP} + \text{Enzyme.}$$

2. Attachment of mRNA. mRNA attaches itself to smaller subunit of ribosome in the region of tunnel. Its cap has nucleotides complementary to the nucleotides present on rRNA. The combination is such that initiation codon (AUG) on mRNA comes to lie at P-site (peptidyl transfer side on the ribosome, which is jointly contributed by the two ribosomal subunits). A protein factor called **initiation factor-3** is required in this process.

3. Initiation of Polypeptide Chain (Fig. 5.12). The synthesis of protein or polypeptide chain starts with amino acid methionine and, therefore, methionyl tRNA ($tRNA_f^{met}$) charged with methionine, binds to the initiator codon on mRNA at the P site. Since the codon for methionine is AUG, the $tRNA_f^{met}$ would have UAC at the anticodon site. The anticodon of ($tRNA_f^{met}$) establishes temporary hydrogen bonds with the initiator codon on mRNA. This process occurs in presence of **initiation factor-2.** It also requires energy which is provided by GTP. The initiation factors involved in attachment and initiation processes are designated as **IFs** in prokaryotes and **eIFs** in eukaryotes.

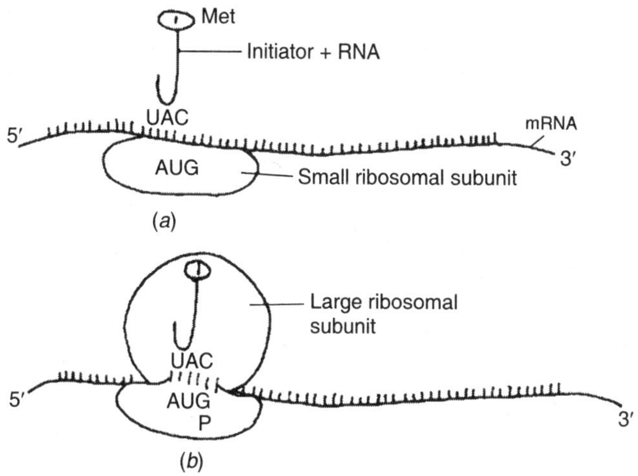

Fig. 5.12. Steps in protein synthesis; (*a*) Initiation: The small ribosomal subunit binds to the mRNA in presence of initiation factors. The interaction is facilitated by a ribosomal RNA recognition sequence on the messenger RNA. At the same time, the initiator methionyl-tRNA interacts with the messenger RNA through codon-anticodon interaction and (*b*) The large ribosomal subunit binds and an initiation complex is thus formed. The initiator methonyl-tRNA is located at the P-site on the ribosome.

The amino acid is linked to the tRNA through its carboxyl group and the $-NH_2$ group of the first amino acid (methionine) is formylated in prokaryotic systems. This would block the $-NH_2$ group of the first amino acid from getting into peptide bond formation. The tRNA used to introduce methionine at the initiation site ($tRNA_f^{met}$) is different from the one that inserts methionine elsewhere in the protein chain. With the initiator tRNA at the P-site, the large ribosomal subunit now attaches to the complex and the intact ribosome (active ribosome) is formed.

4. Elongation (Polypeptide chain formation; Fig. 5.13). A second tRNA charged with an appropriate amino acid enters the ribosome at the A-site. Its anticodon binds to the complementary codon of mRNA chain by hydrogen bonds. A peptide bond is formed between the first amino acid (methionine) and the second amino acid. The reaction is catalysed by the enzyme **peptidyl transferase.** The 23S rRNA of large subunit has been assigned this catalytic function. In this process, the linkage between methionine and its tRNA is broken and the —COOH group now forms a peptide with the free $—NH_2$ group of the second amino acid. Thus, the second tRNA bound at the A-site (amino acyl or acceptor site, situated on the larger subunit of the ribosome and faces the tunnel between the two subunit.) carries a dipeptide (a peptide formed with two amino acids) and the first methionyl tRNA is removed from the P-site. Now, the tRNA at the A-site is pulled to the P-site along with the mRNA. This process is called **translocation.** All these steps require GTP and a number of proteins called **Elongation Factors** (**EFs**). In prokaryotes, elongation factors are called **EF-Tu, EF-Ts** and **EF-G.** Eukaryotes require a more complex set of accessory factors. The GTP is hydrolysed to GDP and inorganic phosphate to release energy for the process. As a result of translocation the third codon comes into the A-site and an appropriate tRNA charged with a third amino acid would bind at the A-site. The process of peptide bond formation and translocation would be repeated. Thus, as the mRNA moves with respect to the ribosome, all the codons would be expressed at the A-site and the peptide chain would grow. Thus, the amino acids are linked up into a polypeptide in a sequence communicated by the DNA through the mRNA. The four letter language of DNA (A, T, G, C) is transcribed into a four letter language of mRNA (A, U, G, C), which is translated into a twenty letter language of polypeptides.

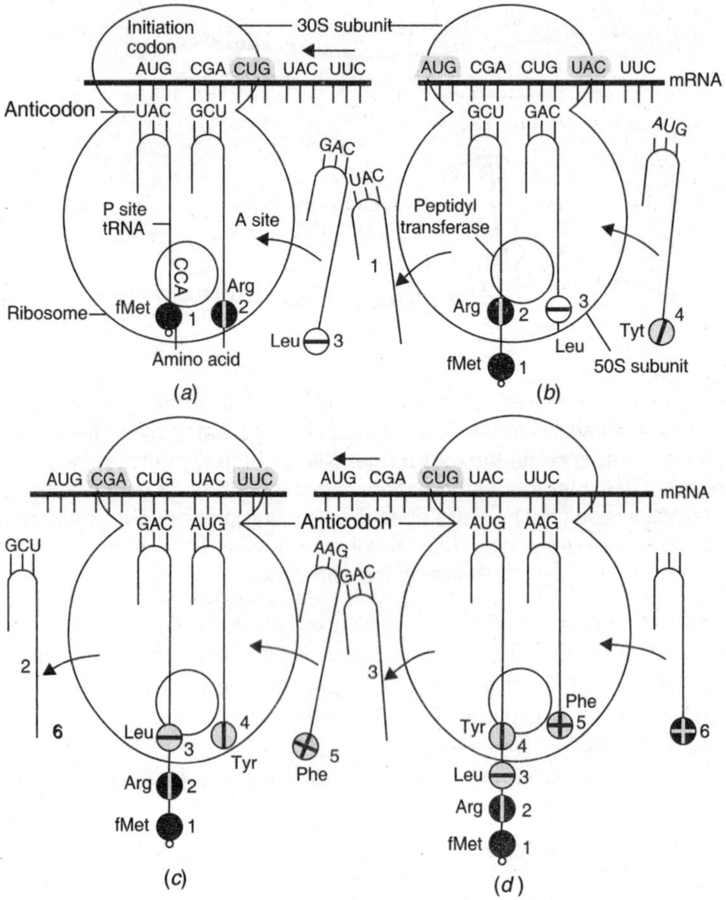

Fig. 5.13. Elongation of polypeptide chain.

5. Termination. Synthesis of polypeptide terminates when a non-sense codon of mRNA reaches the A-site. There are three non-sense codons UAA, UAG and UGA. These codons are not recognised by any of the tRNAs. Therefore, no more amino-acyl tRNA reaches the site. The tRNA at P-site is hydrolysed and the completed polypeptide is released in the presence of GTP dependent release factors (**RF_1, RF_2** and **RF_3**). The ribosome slips off the mRNA chain (Fig. 5.14). The two subunits of ribosome also dissociate in presence of **dissociation factor.** The subunits of ribosome and the tRNA on release from the mRNA can function again in the formation of other polypeptides.

The just released polypeptide is a straight, linear molecule having a primary structure. It may lose some amino acids from the end with the help of **exopeptides enzyme,** and then coil and fold on itself to aquire secondary and tertiary structures. It may combine with other polypeptide to have quaternary structure.

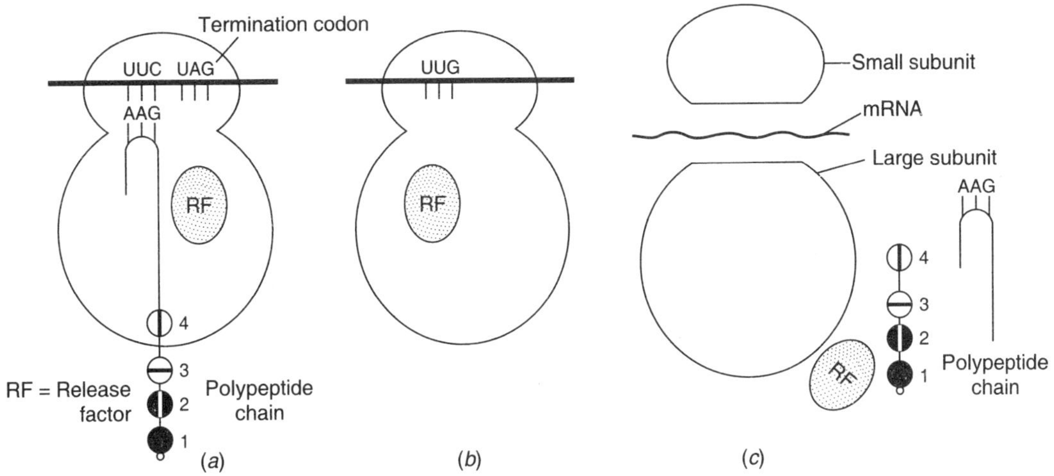

Fig. 5.14. Termination of polypeptide chain.

> **Protein Translocation**
>
> When protein synthesis is terminated, the ribosomal subunits, fall off the mRNA. In case of free cytoplasmic polyribosomes, the polypeptides or proteins are released into the cytoplasm (cytosol). These proteins remain in the cytoplasm and are employed in the synthesis of more cytoplasm and other cellular components. Some specific proteins are translocated into mitochondria and nucleus through special mechanisms. In the case of membrane bound polyribosomes, the polypeptide chain, as it grows on the messenger RNA, is inserted into the lumen to the endoplasmic reticulum. Some of the proteins get integrated into the membrane and thus become integral membrane proteins. Some others are released into the lumen and are incorporated in Golgi vesicles. These proteins undergo other modification such as **glycosylation** (addition of sugar residues). The modified proteins are packed in vesicles for export or formation of lysosomes, cell wall enzymes, plasma membrane, etc. (Fig. 5.15).

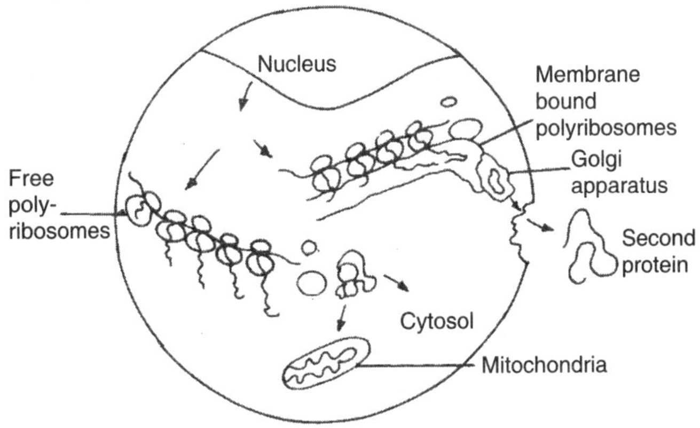

Fig. 5.15. Synthesis of proteins on free and membrane-bound polyribosomes.

REGULATION OF GENE EXPRESSION OR PROTEIN SYNTHESIS

Gene expression refers to the mechanism at the molecular level by which a gene is able to express itself in the phenotype of an organism. The mechanism of gene expression involves biochemical genetics. It consists of synthesis of specific RNAs, polypeptides, structural proteins and enzymes which control the structure or functioning of a specific trait.

The regulation of gene expression is a very broad term, that may occur at various levels. In eukaryotes, the regulation could be exerted at (*i*) transcriptional level (formation of primary transcript), (*ii*) processing level (regulation of splicing). (*iii*) transport of mRNA from nucleus to the cytoplasm level, (N) translational level, etc.

The mechanism which stimulates the expression of certain genes and inhibits that of others is called regulation of gene expression. Though, all the mature cells of a living being possess the same gene complement or genetic material, even then they differ in their structure and functioning. It is possible only if the organism has a mechanism of regulating gene activity by allowing some to function and others to restrain their activity through a switching on and switching off system. This means, the genes are turned 'on' or 'off' as per requirement. A set of genes is 'switched on' when there is necessity to handle and metabolise a new substrate. The enzymes produced by these genes metabolise the substrate. The molecules of metabolite iliciting the switching on of the genes are termed as **inducers** and the phenomenon is called **induction.** Similarly, certain genes which are in their 'switch on' state, continue to synthesise a metabolite till the later is produced in amount more than required or else it is supplied to the cell from outside. In other words, certain genes continue to express themselves till the end product of their enzyme activity inhibits or repress their expression. Inhibition by an end product is known as **'feedback repression'.**

Operon Concept

In 1961, two French microbiologist **Francis Jocob** and **Jacques Monad** at the Pasteur Institute in Peris, proposed a mechanism called **operon model** for the regulation of gene action in *E. coli. An operon is a part of genetic material or DNA, which acts as a single regulated unit having one or more structural genes—an operator gene, a promoter gene, a regulator gene, a repressor and an inducer or co-repressor (from outside).* Operons are of two types—**inducible** and **repressible.** An inducible operon system normally remains in switched off condition and begins to work only when the substance to be metabolised by it is present in the cell. Inducile operon system generally occur in catabolic pathways. *e.g.,* Lac operon of *E. coli.* A repressible operon system is normally in its switch on state and continue to synthesise a metabolite till the latter is produced in amount more than required, or else it becomes available to the cell from outside. Repressible operon system is commonly found in anabolic pathway, *e.g.,* Tryptophan operon of *E. coli.* Some other examples of operon are 'ara operon', 'his operon', 'val operon', etc.

1. Inducible System (Lac operon of *E. coli*)

An inducible operon system consists of four types of genes and chemicals (*Fig. 5.16*) described below :

(*i*) ***Structural Genes.*** These genes synthesise mRNAs, which in turn synthesise polypeptide or enzyme over the ribosomes. An operon may have one or more structural genes. Each structural gene of an operon is called **cistron.** The lac operon (lactose operon) of *Escherichia coli* contains three structural genes (Z, Y and A). These genes occur adjacent to each other and thus are linked. They transcribe a polycistronic mRNA molecule (a single stretch of mRNA covering all the three genes), that helps in the synthesis of three enzymes-β-galactosidase, lactose permease and transacetylase.

(*ii*) ***Operator Gene.*** It lies adjacent to the structural genes and directly controls the synthesis of mRNA over the structural genes. It is switched off by the presence of a repressor.

An inducer can take away the repressor and switch on the gene that directs the structural genes to transcribe.

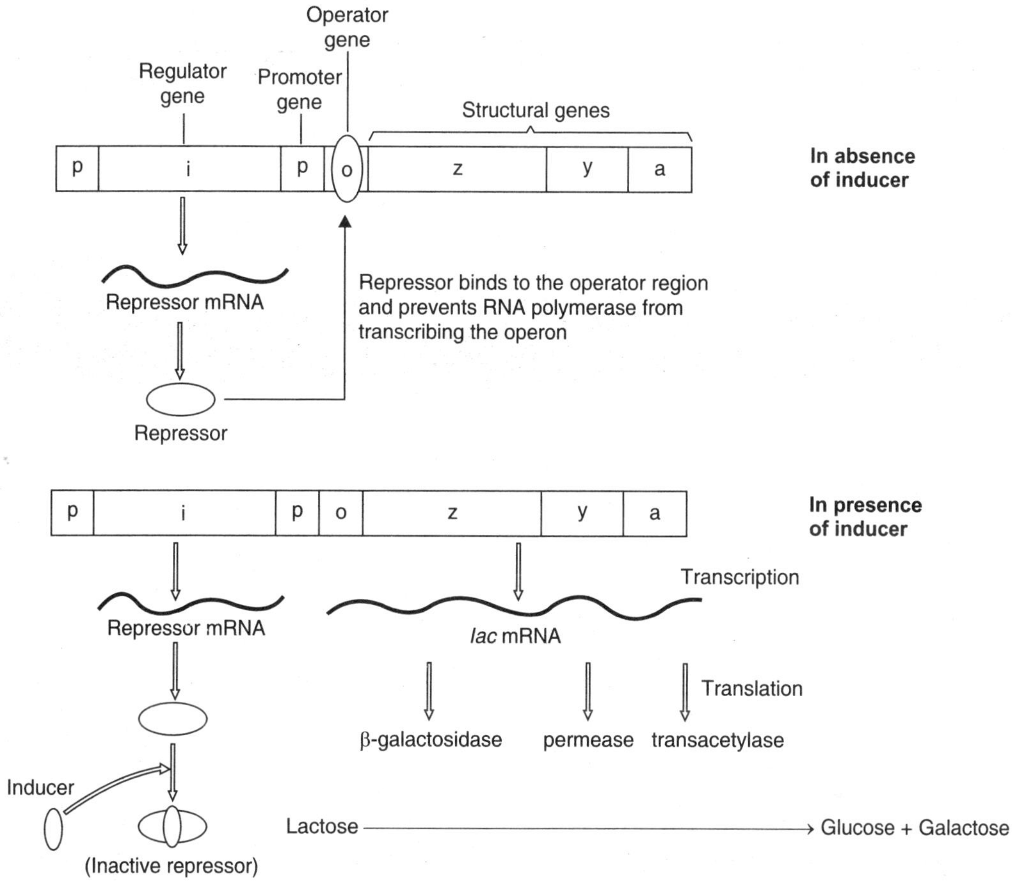

Fig. 5.16. The *lac* Operon.

(*iii*) ***Promoter Gene.*** This gene is the site for initial binding of RNA polymerase. When the operator gene is turned on, the enzyme RNA polymerase moves over it and reaches the structural genes to perform transcription of mRNA.

(*iv*) ***Regulator Gene.*** It produces a repressor that binds to operator gene and stops the working of the latter. A regulator gene is not necessarily adjacent to the operator gene it controls. Here gene *i* does not refer to inducer, rather it is derived from the word inhibitor.

(*v*) ***Repressor.*** It is a protein produced by the regulator gene. It binds to the operator gene so that the transcription of mRNA by structural gene stops.

(*vi*) ***Inducer.*** It is a chemical (substrate, hormone or some other metabolite) which after coming in contact with the repressor, forms an **inducer repressor complex.** This complex cannot bind with the operator gene, which is thus switched on. The freed operator gene allows the structural gene to transcribe mRNA to synthesise the enzymes.

The inducer for lac operon of *Escherichia coli* is lactose (in fact allolactose an isomer of lactose). When the sugar lactose is added to the culture of *E. coli*, a few molecules of lactose gets into the bacterial cells by the action of the enzyme permease, a small amount of this enzyme is present in the cell even when the operon is not working. These few lactose molecules are then converted into an active form which acts as an inducer and binds to the repressor, a product of regulator gene. The inducer repressor complex fails to join with the operator, which is turned on. The three genes are expressed as three enzymes to metabolise lactose.

2. Repressible System (Tryptophan operon of *E. coli.*)

Tryptophan operon of *Escherichia coli* is an example of repressible system (Fig. 5.17). It consists of the following :

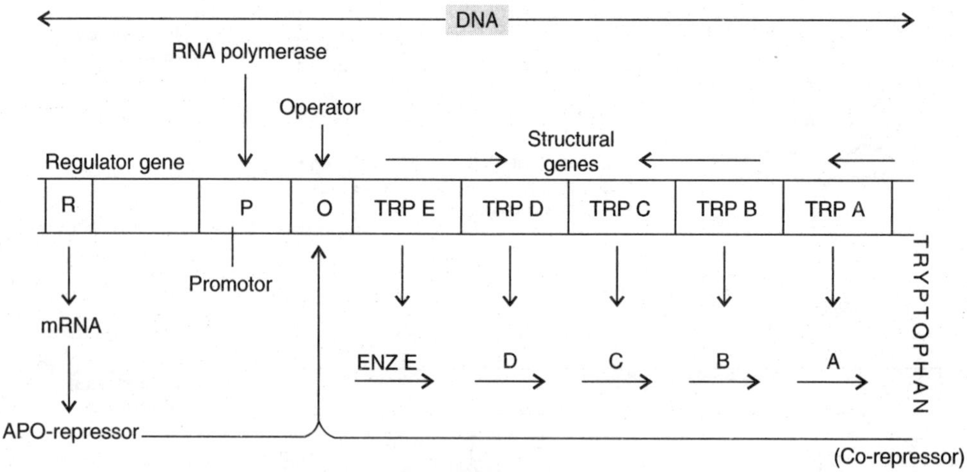

Fig. 5.17. A repressible tryptophan operon.

(*i*) ***Structural Genes.*** These genes are meant for transcription of mRNA, which in turn synthesise polypeptide or enzyme. Tryptophan operon has five structural genes E, D, C, B and A. They lie in continuation and synthesise enzymes for five steps of tryptophan synthesis.

(*ii*) ***Operator Gene.*** It lies adjacent to the structural genes and controls the functioning of the latter. Normally it is kept switched on, because the apo-repressor produced by the regulator gene does not binds to it. The operator gene is switched off when a co-repressor is available along with apo-repressor.

(*iii*) ***Promoter Gene.*** It marks the site at which the RNA polymerase enzyme binds. When the operator gene is switched on, it moves from promotor gene to structural genes for transcription of mRNAs.

(*iv*) ***Regulator Gene.*** It produces a regulatory protein called apo-repressor for possible blocking the activity of operator gene.

(*v*) ***Apo-repressor.*** It is a regulatory protein synthesised by regulator gene. When a co-repressor substrate is available in the cell, the apo-repressor combines with the co-repressor to form an apo-repressor-co-repressor complex. This complex binds with the operator gene and switches it off. When the apo-repressor is not available in proper strength, the operator gene is kept switched on because, by itself the apo-repressor is unable to block the working of operator gene.

(*vi*) ***Co-repressor.*** It is an end product of reactions catalysed by enzymes produced by the structural genes. It is usually a non-protein component of apo-repressor. The end product rarely accumulates because it is often utilised in some other reaction and hence does not function as co-repressor. However, whenever it accumulates and become available to the cell from outside, it combines with the apo-repressor to form apo-repressor-co-repressor complex and blocks the operator gene. In case of tryptophan operon, the amino acid tryptophan functions as a co-repressor.

Differences between Induction and Repression

Induction	*Repression*
1. It is switching-on of an operon which is normally in switched-off state.	1. It is switching-off an operon which is normally in switched-on state.
2. It starts transcription and translation.	2. It stops transcription and translation.
3. It is caused by a new metabolite which needs enzymes to get metabolised.	3. It is caused by increased formation or availability of a metabolite.
4. It generally operates in a catabolic pathway.	4. It generally operates in an anabolic pathway.
5. It removes the repressor of an operon by the inducer metabolite.	5. It is the blocking of the operator gene of an operon through a complex formed by the union of apo-repressor formed by a regulator gene and co-repressor which is actually a product of anabolic pathway.

IMPORTANCE OF GENE REGULATION

(*i*) The constitutive genes are not associated with regulator and operator genes and therefore, operate all the time because their products are constantly needed for cellular activity.

(*ii*) In regulated gene action, all the genes required for a multistep reaction can be switched on or off simultaneously.

(*iii*) The genes are switched on or off in response to particular chemicals whether required for metabolism or are formed at the end of a metabolic pathway.

(*iv*) Gene regulation is required for growth, division, differentiation and morphogenosis.

REVIEW EXERCISES

MULTIPLE CHOICE TYPE QUESTIONS

1. The most abundant organic components of the cells are:
(*a*) Lipids (*b*) Proteins
(*c*) Carbohydrates (*d*) Nucleic acids.

2. Which of the following is a storage protein?
(*a*) Keratin (*b*) Collagen
(*c*) Glutelin (*d*) Haemoglobin.

3. The Most diverse molecules in a cell are
(*a*) Mineral salts (*b*) Lipids
(*c*) Carbohydrates (*d*) Proteins.

4. Which of the following is a fibrous protein?
(*a*) Collagen (*b*) Globulin
(*c*) Haemoglobin (*d*) Hordein.

5. mRNA directs the building of proteins through a sequence of
(*a*) Introns (*b*) Codons
(*c*) Exons (*d*) Anticodon.

6. The flow of information from DNA to mRNA and then to protein is called
(*a*) Transcription (*b*) Translation
(*c*) Central dogma (*d*) Genetic code.

7. Transcription begins when one of the following enzymes binds to a promotor site.
(*a*) DNA polymerase (*b*) RNA polymerase
(*c*) Helicase (*d*) Gyrase.

8. Elongation of RNA polynucleotide chain always takes place
(*a*) when RNA acts as a primer
(*b*) $5' \rightarrow 3'$ direction with new nucleotide always added at 3' end
(*c*) during replication of DNA
(*d*) in any direction along the entire length of single DNA strand.

9. Amino-acyl synthetase enzyme takes part in
(*a*) attachment of mRNA of 30s ribosome
(*b*) activation of amino acid
(*c*) transfer of activated amino acid to tRNA
(*d*) Hydrolysis of ATP to AMP.

10. If the sequence of bases in DNA is ATTCGATG, then the sequence of bases in its transcript will be
(*a*) GUAGCUUA (*b*) UAAGCUAC
(*c*) CAUCGAAU (*d*) AUUCGAUG.

11. Initiation codon of protein synthesis (in eukaryotes) is
(*a*) GUA (*b*) GCA
(*c*) CCA (*b*) AUG.

12. The process of translation is
(*a*) Ribosome synthesis (*b*) Protein synthesis
(*c*) DNA synthesis (*d*) RNA synthesis.

13. Which is not involved in protein synthesis?
(*a*) Transcription (*b*) Initiation
(*c*) Elongation (*d*) Termination.

14. Central dogma was proposed by
(*a*) Beadle and Tatum (*b*) Temin and Baltimore
(*c*) Klung (*d*) Crick.

15. During transcription, the DNA site at which RNA polymerase binds is called
(*a*) Enhancer (*b*) Promoter
(*c*) Regulator (*d*) Receptor.

OBJECTIVE TYPE QUESTIONS

1. Fill in the blanks:
(*i*) A protein molecule is a polymer of ....
(*ii*) Bond —CO—NH— established between two amino acid is called .... bond.
(*iii*) Keratin is a .... protein.
(*iv*) Transfer of information from DNA to mRNA is called ....
(*v*) A sequence of 3 bases that code for an amino acid is called a ....
(*vi*) The formation of peptide bond is catalysed by the enzyme ....

2. Mark the statements True (T) and False (F)
(*i*) The synthesis of RNA molecule from DNA template is called translation.
(*ii*) Protein synthesis involves three steps: initiation, elongation and termination of polypeptide chain.

(*iii*) Assembly of an amino acid chain according to the sequence of base triplets in mRNA is called transcription.

(*iv*) Collagen is a structural protein found in connective tissues of animals.

(*v*) Myoglobin is a protein involved in transport of oxygen to the tissues.

TEST QUESTIONS

1. What is a polypeptide? How is it formed?
2. List four functions of proteins and name one protein that performs each function.
3. Describe biological significance of proteins.
4. Why hnRNA is required to undergo splicing?
5. Describe the two additional processings which hnRNA needs to undergo after splicing so as to become functional.
6. Mention the role of the codon AUG and UGA during protein synthesis.
7. How is protein synthesis is initiated in a cell?
8. How is elongation carried on during protein synthesis in a cell?
9. Write full names of different types of RNA. State only, how each type is involved in protein synthesis?
10. Write about tRNA as follows:
    (*i*) Characteristics of its structure.
    (*ii*) Part of the cell where they are located.
    (*iii*) Technical name of the set of bases that are complementary to the triplet code on mRNA.
11. Differentiate between
    (*i*) Codon and anticodon
    (*ii*) Transcription and translation.
12. Enlist the role of following in protein synthesis
    (*i*) mRNA (*ii*) rRNA
    (*iii*) tRNA (*iv*) ribosomes
    (*v*) Amino acids (*vi*) ATP.
13. Describe an inducible operon and differentiate it from a repressible operon.
14. What is an operon? Name the three types of genes which make up an operon.
15. Describe the process of protein synthesis with reference to initiation, elongation and termination.
16. Discuss the characteristics of genetic code.

UNIT–3

CELL PROLIFERATION

6 Cell Division

CONCEPT OF REPRODUCTION

Living organisms do not live forever. There comes a time, when an organism dies. Death can be caused by a number of factors such as diseases, accidents, predation, starvation or pollution. If organisms do not die due to such factors, they eventually die because of old age. Death of every individual organism is a certainty (*i.e.*, no individual immortal), still life continues. There must be some process in living organisms that ensures this continuity. The process that ensures continuity of life is referred as reproduction. It is necessary for the continuation of life and to compensate for the loss of life due to death. The statement '*nothing lives forever yet continues*', explain the role of reproduction in nature.

Reproduction is defined as 'a biological process in which an organism gives rise to young ones (offspring) similar to itself. The offspring grow, mature and in turn produce new off-spring. Thus, there is a cycle of birth, growth and death. Reproduction enables the continuity of a species generation after generation.

Types of Reproduction

Based on whether there is participation of one organism or two in the process of reproduction, it is of two types: asexual and sexual.

1. **Asexual Reproduction.** When offspring is produced by a single parent and without the involvement of gamete formation, the reproduction is called asexual reproduction. The offspring produced through asexual reproduction are not only identical to one another but are also exact copies of their parent. A group of morphologically and genetically similar individuals is called **clone.**

Asexual reproduction is common among single-called organisms, and in plants and animals with relatively simple organisations. In asexual reproduction it is necessary for one cell to divide into two genetically identical cells. The cell division that takes place in the body cells (also called **somatic cells**) and produce identical cells is known as **mitosis.** Asexual reproduction is characterised by :

- involvement of single parent (uniparental condition).
- absence of formation of gametes and their fusion (fertilisation).
- involves only mitotic cell division.
- daughter individuals are genetically identical to the parent.
- multiplication/formation of new individuals is quite rapid.

2. **Sexual Reproduction.** Sexual reproduction involves formation of the male and female gametes either by the same individual or by different individuals of the opposite sex. These gametes fuse to form the zygote which develops to form the new organism. It is an elaborate, complex and slow process as compared to asexual reproduction. Because of the fusion of male and female gametes, sexual reproduction results in offsprings that are not identical to the parents or amongst themselves.

Sexual reproduction is characterised by :

- Usually involves two parents (biparental).
- involves formation of gametes and their fusion (fertilisation).
- involves both meiotic and mitotic cell divisions.
- the daughter individuals genetically differ from the parents.
- formation of new individuals is a slow process.

CELL CYCLE AND CELL DIVISION

All multicellular organisms start life as a single cell—the fertilized egg, and grow by addition of new cells. The new cells arise by the division of pre-existing cells. This was first suggested by **Rudolf virchow** (1958) as *'Omnis cellula e cellula'* meaning every cell is derived from a cell. **Strasburger** (1873), similarly proposed that, nuclei are formed from pre-existing ones. The process of formation of new cells by the division of pre-existing cells is called **'cell reproduction'. Flemming** (1880) studied the details of somatic cell division and coined the term **'mitosis'** for it. **Van Benedin** (1887), **Strasburger** (1880), **Sutton** (1990) and **Winiwater** (1900) studied the details of division that occur prior to the formation of gametes. They found that the number of chromosomes is reduced to half in gametes as compared to their parent cells. The division is termed as **'meiosis'** by **Former** and **Moore** (1905).

The mode of cell division is fundamentally similar in all organisms. This emphasises the unity of life. A cell divides when it has grown to a certain maximum size, which disturbs the **karyoplasmic** or **kern plasma ratio.** Cell division is also controlled by a certain agent, factor or substance that triggers cell division. It is called **mitogen**. The common plant mitogen is a hormone cytokinin. The common animal mitogen are lymphokines, Epidermal Growth Factor (EGF), Platelet Derived Growth Factor (PDGF) etc. Daughter cells formed after the division of a parent cell may grow in size and divide again. The period between two successive divisions is called **generation time**. It varies from a few minute to a few days, depending upon the type of cell and its environmental conditions.

Cell Cycle

The sequence of events, by which a cell duplicates its genome and synthesizes other cell contents and eventually divides into two daughter cells is termed as **cell cycle.** A typical eukaryotic cell cycle (*e.g.,* human cells in culture) divide approximately every 24 hours. However, this time span can vary from organism to organism and also from cell to cell type. For example, in yeast, cell cycle takes only about 90 minutes. Cell cycle consists of two stages, a non-dividing growing **interphase** or **I-phase** and a short dividing **mitotic** or **M-phase** (Fig. 6.1)

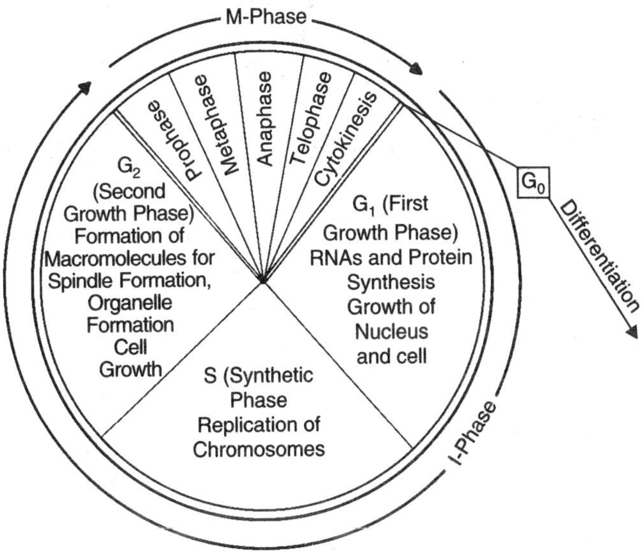

Fig. 6.1. Cell cycle.

I. Interphase (I-Phase)

Interphase (L. *inter*—between; Gk. *phasis*—aspect) is a series of changes that occur in a newly formed cell and its nucleus before it becomes capable of division again. Therefore, it is also called **intermitosis.** It is a non-dividing state of the cell and its nucleus is called **interphase nucleus.** Previously this phase was called "resting stage" but is in fact a period of great activity. Three important processes, which are preparatory to cell division take place during interphase. These processes are : (1) Replication of DNA along with the synthesis of basic nuclear proteins, the histones. (2) In animal cells, the centriole divides to form a pair of new centrioles. Duplication of a centriole takes place by the outgrowth of daughter centrioles from the parent centrioles, which are at right angle to each other. (3) Synthesis of energy-rich compounds, which provide energy for mitosis, and synthesis of proteins (which are later found in the spindle) at the end of interphase.

Interphase takes approximately 95 per cent of the total duration of the cell cycle and lasts for 10–20 hours. Whereas, the dividing phase takes only about an hour. Interphase is further divisible into three stages—G_1-phase, S-phase and G_2-phase.

1. **G_1-Phase.** It is also called **first growth phase** or **post mitotic gap phase.** During this phase, the cell grows in size and there is an active synthesis of RNA and proteins. In this phase the cell carries out its physiological functions and prepares the machinery needed for the cell to proceed to the next stage. A large number of nucleotides, amino acids for histone

synthesis and enregy rich compounds are formed. Cell organelles also increase in number. However, it shows no change in its DNA content.

2. **S-Phase.** It is also called **synthetic phase.** In this phase, DNA molecule of each chromosome replicates by the synthesis of a new DNA on the template of the existing DNA. Thus, the DNA content doubles and duplicate set of genes are formed. Along with DNAs, chromatin fibres also replicate. As chromatin fibres are elongated chromosomes, each chromosome comes to have two chromatin threads or sister chromatids which remain attached at a common point called centromere. The cell thus retains the original dipliod ($2n$) chromosome number but now has a duplicate set of genes. S-phase is also called invisible phase of mitosis, since in this phase chromosome prepare themselves for equitable distribution later on. The **centriole,** also divides into two centriole pairs in the cells containing the same.

3. **G_2-phase.** It is also called **second growth phase** or **premitotic gap phase.** The synthesis of RNA and protein continues in this phase and the cell prepares itself to go into the mitotic phase. The phase produces macro molecules for multiplication of cell organelles, spindle formation and cell growth. During G_2-phase a cell contains double DNA content *i.e.,* 1C to 2C for haploid cells and 2C to 4C for diploid cells.

II. Mitotic Phase (M-phase)

Mitotic phase or M-phase follows the interphase and represents the phase of actual division. It consists of **karyokinesis** (division of nucleus) and **cytokinesis** (division of cytoplasm). After M-phase a cell may either enter interphase to repeat the cell cycle or **G_0-phase (quiescent stage)** to arrest the cell cycle (due to non-availability of mitogens and energy rich compounds). Thus, G_0-phase stops further division, and the cell may grow in size and get differentiated. They assume a particular shape and come to have a particular function. This phenomenon is called **cell differentiation.**

MODES OF CELL DIVISION

Cell division occurs in two ways : mitosis and meiosis. In each case, division of the nucleus occurs before the division of the cytoplasm.

Mitosis : Cell Replication

Mitosis (Gk. *mitos*—thread of fibril) *is the process of nuclear division in which chromosomes replicate and become equally distributed into two daughters so that the daughter cells come to have the same number and type of chromosomes as are present in the parent cell.* Mitosis was first observed by **Strasburger** (1870) in plant cells and later by **Boveri** and **Flemming** (1879) in animal cells. The term 'mitosis' was coined by **Flemming** (1882).

Mitosis is the common method of cell division which brings about growth in multicellular organisms and multiplication in unicellular organisms. It occurs in the formation of somatic body cells and is, therefore, often called **somatic cell division.** Mitosis keeps the chromosome number in the daughter cells equal to the parent cell. Hence, it is also known as the **equational division.**

The common sites of mitosis cell division in plants are meristematic regions, *e.g.,* stem tip, root tip, embryo, leaves, flowers, fruits, etc. In animals, mitosis occurs in skin epidermis, bone marrow and embryo. Depending upon the type of cells, mitosis lasts from 30 minutes to 3 hours.

Mitosis consists of two main events—karyokinesis and cytokinesis.

Karyokinesis

Karyokinesis (Gk. *karyon*-nucleus ; *kinesis*-movement) is also called indirect nuclear division, because the nucleus passes through a series of uninterrupted changes before forming two daughter nuclei. Though, karyokinesis is a continuous process, it has been divided into four phases for the sake of convenience. They are prophase, metaphase, anaphase and telophase (Figs. 6.2 and 6.3).

1. **Prophase** (Gk. *pro*—first ; *phasis*—stage). In early prophase, the chromatin fibres shorten and thicken to form slender thread like structures called **chromosomes.** Therefore, chromatin network disappears. The shortening and thickening of chromatin threads occur due to coiling or spiralisation of chromatin fibres and loss of water by proteins surrounding the chromatin fibres. The elongated chromosomes may show overlapping. At this stage the nucleus looks like a loose ball of wool. Because of this appearance, the prophase was formerly termed as **spireme stage.**

In the beginning of prophase, in an animal cell, the already duplicated centrosome or centriole pair begin to shift towards the opposite sides. Both the centriole pairs radiate out fine microtubular fibrils called astral rays, from their pericentriolar satellites. Each group of astral rays along with its centriole pair is called **aster.** The two asters later move to opposite poles and initiate spindle formation.

In middle prophase, the chromosome become peripheral. They shorten and thicken further to assume characteristic shape and size. The shortening of chromosomes ensures their equitable distribution later during anaphase. Each chromosome appears to consist of two longitudinal identical threads called sister chromatids. The two chromatids of a chromosome are attached to each other by means of a narrow point called **primary constriction** or **centromere.** At this region, each chromatid has a disc like structure, the **kinetochore,** where the spindle fibres join it. Nucleolus or nucleoli, found attached to one or more chromosomes start decreasing in size.

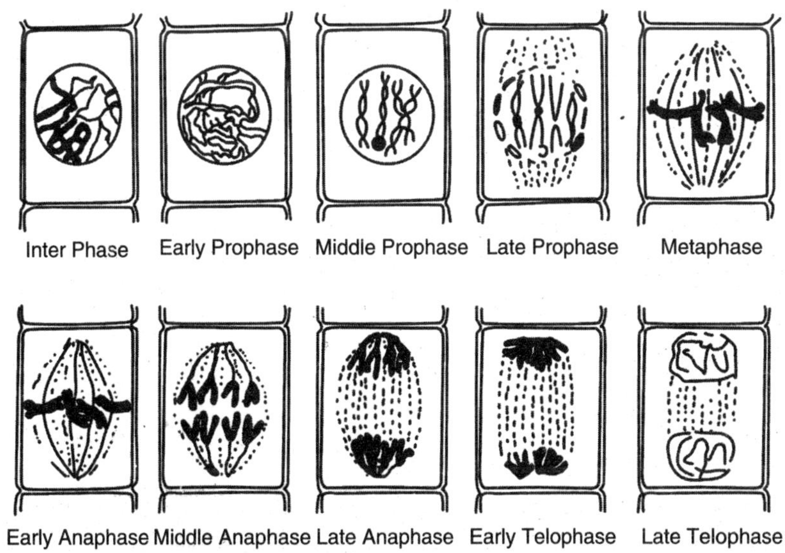

Fig. 6.2. Different stages of mitosis in a plant cell.

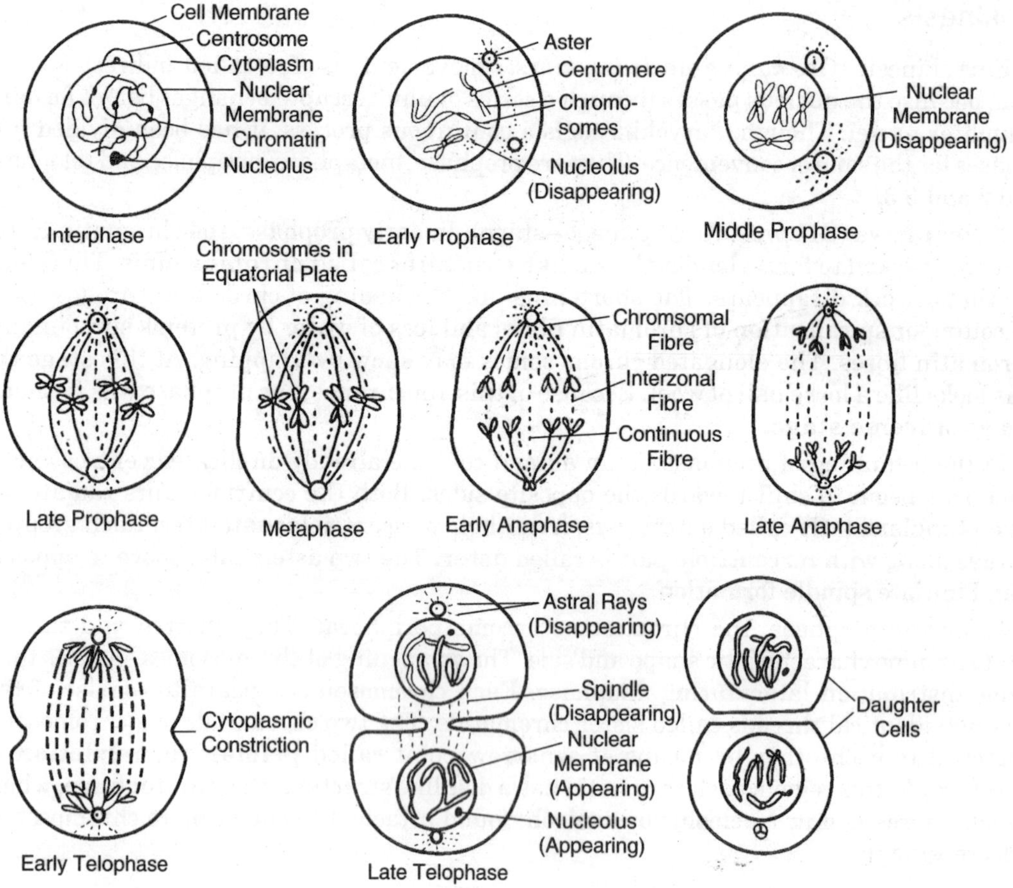

Fig. 6.3. Different stages of mitosis in an animal cell.

In the late prophase (**prometaphase**) the nuclear envelope breaks up into small vesicles, simultaneously, nucleolus diminishes in size and gradually disappear. But this time the two asters come to lie in the area of future spindle poles.

2. **Metaphase** (Gk. *meta*—after or second ; *phasis*—stage). The disintegration of nuclear envelope results in the formation of a fibrous body all over the nucleoplasm. It is called **spindle, acromatic spindle** or **spindle apparatus.** It consists of fine fibres called microtubules that are made up of proteins, **tubulins.** The spindle fibres converge towards the two ends called **poles.** In animal cells, the asters act as poles. Since there is an aster at each pole, the spindle of animal cells is called **amphiastral** (Gk. *amphi*—both ; *aster*—star). In plants, there is no centrioles and no asters are formed, therefore, the spindle of plant cells is called **anastral** (Gk. *an*—not ; *aster*—star). The spindle apparatus has maximum diameter in the middle called **equator.** The asters that surround the centriole and the spindle together constitute the **mitotic apparatus.**

Chemically the spindle apparatus is composed of 95–97 per cent protein tubulin, 3.5–5 per cent RNA and traces of lipids and other substances. Some of the spindle fibres run from pole to pole. They are called **continuous** or **interpolar fibres.** Others radiate out from one pole but do not reach the other. They get attached to the two centromere surfaces (kinetrochores) of each chromosomes. They are called **discontinuous fibres, chromosomes fibres** or **tractile**

fibrils. Each chromosome is attached to both the spindle poles by distinct chromosome fibres (Fig. 6.4).

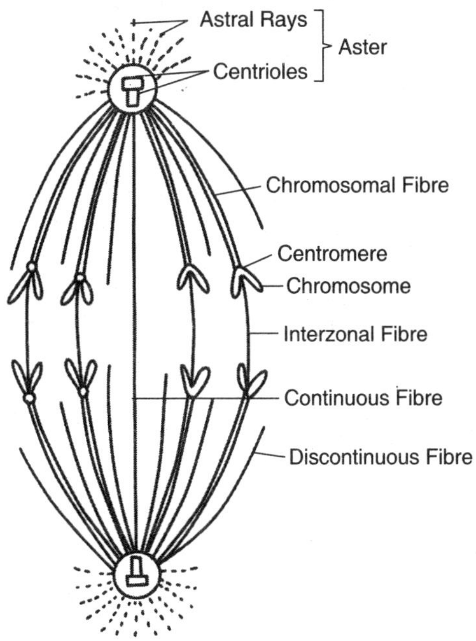

Fig. 6.4. Chromosome asters and spindle in mitotic anaphase.

The chromosome fibres now tighten and drawn the chromosomes at the equator of spindle. The process of drawing the chromosomes on to the equator of the spindle is known as **congression.** The chromosomes get arranged in the form of a plate, called the **equatorial** or **metaphase plate.** This plate is at right angles to the axis of spindle, and is actually formed of the kinetochores, the arms of chromatids trailing away. On the equator, the small chromosomes are towards the interior, while the larger ones are towards the periphery. The events which connect the chromosome to the spindle fibres and bring them to the metaphase plate are sometimes referred as **prometaphase.** The nuclear division in which nuclear membrane dissolves and there is no distinction between cytoplasm and nucleoplasm, is referred as **extra-nuclear mitosis** or **eumitosis.** In many protists, fungi and algae nuclear envelope does not degenerate, spindle is formed inside the nucleus and mitosis occurs there. It is called **premitosis** or **intranuclear mitosis.** It may be centric or acentric.

Metaphase is the most suitable time to count the number of chromosomes and study their morphology as the chromosomes are most condensed in this phase.

3. **Anaphase** (Gk. *ana*—up ; *phasis*—stage). The sister chromatids of each chromosome separate completely to become daughter chromosomes (Fig. 6.5). The daughter or new chromosomes move towards the poles along the path of their chromosome fibres or tractile fibrils. During anaphasic movement of chromosomes, the centromeres lead the path and the arms trail behind. As a result, the anaphasic chromosomes appear V, L, J and I shaped. The shapes are formed respectively in metacentric, sub-metacentric, acrocentric and telocentric chromosomes.

It is held that the anaphasic movement of chromosomes is caused by the (*i*) contraction of chromosome fibres, (*ii*) formation and expansion of interzonal fibres and (*iii*) poleward dissolution of microtubules of chromosome fibres.

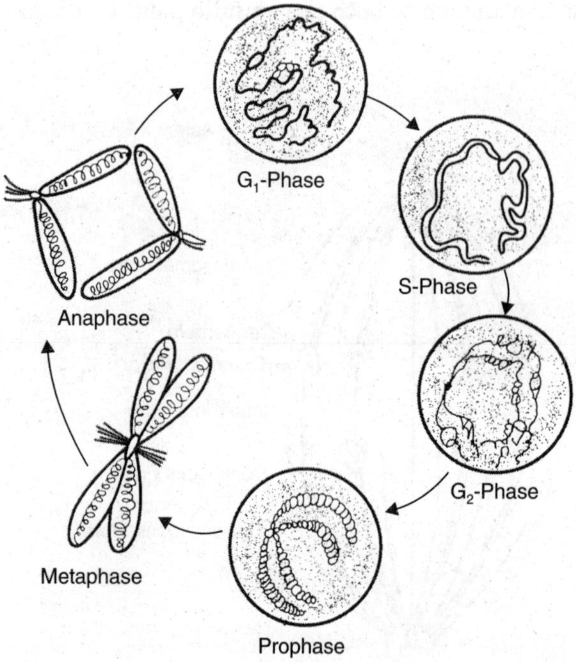

Fig. 6.5. Behaviour of a chromosome during interphase and mitosis.

At the end of anaphase, there are two similar groups of chromosomes at each pole. The tractile fibrils disappear, while the spindle apparatus persists for sometime. The number of chromosomes at each pole remains the same as in the parent nucleus.

4. **Telophase.** (Gk. *telos*—end ; *phasis*—stage). This phase begins with the arrival of daughter chromosomes at the two poles. The events of telophase are reverse to those of the prophase. The chromosomes despiralise and become thin, granular and thread like. These threads overlap one another to form a fine chromatin network. Nucleoli are formed at the nucleolus organising regions of nucleolar or satellite chromosomes. Nucleoplasm collects in the area of chromatin. Nuclear envelope reappears around each group of chromosomes, either from pieces of the older nuclear envelope or endoplasmic reticulum. Thus, the nucleus of each daughter cell is reconstituted.

Cytokinesis

Cytokinesis (Gk. *kytos*—hollow or all ; *kinesis*—movement) is the division of cytoplasm of the parent cell into two daughter cells. Normally it starts towards the middle anaphase and is completed simultaneously with the telophase. Cytokinesis occurs differently in animal and plant cells.

Cytokinesis in Animal Cells

Animal cells typically divide by furrowing also called **cleavage cytokinesis.** The central equatorial part of spindle gets changed into dense fibrous and vascular structure called **midbody.** Simultaneously, microfilaments collect in the middle region of the cell below the cell membrane. They include the cell membrane to invaginate. The furrow deepens centripetally and finally divide the parent cells into two daughter cells. As the furrow deepens, the midbody is compressed and becomes smaller and finally disappears as the two cells are fully formed (Fig. 6.6).

Cleavage cytokinesis also occurs in some lower organisms like bacteria, fungi and some algae.

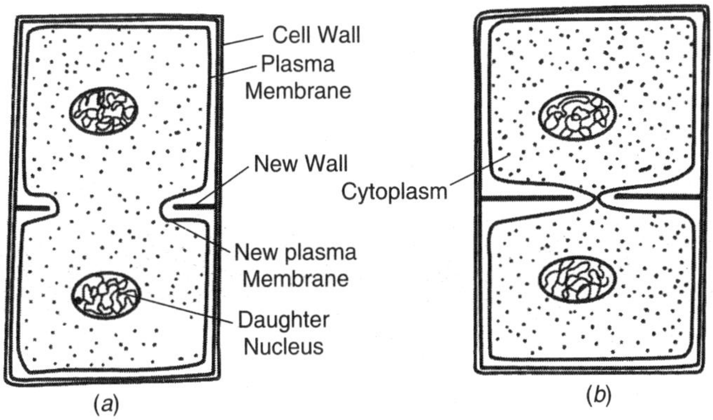

Fig. 6.6. Cytokinesis in bacteria, algae and fungi.

Cytokinesis in Plant Cells

A plant cell cannot undergo cytokinesis by an invaginating cleavage furrow, due to the presence of a rigid cell wall. Therefore, plant cells divide by **cell plate formation** (Fig. 6.7). In this case spindle persists for sometime. It is termed as **phragmoplast.** Small vesicles produced by Golgi apparatus line up at the equator of the phragmoplast. The membranes of vesicles fuse to form the plasma membranes of the two adjacent daughter cells and their contents form a matrix or film. Soon the film solidified to form **cell plate** or **middle lamella.** It grows centrifugally and comes in contact with lateral walls of parent cell. With the formation of the cell plate the spindle or phragmoplast disappears. Cellulose, hemicellulose and pectin are then deposited by the cytoplasm of the daughter cells on either side of the middle lamella to form the primary cell wall.

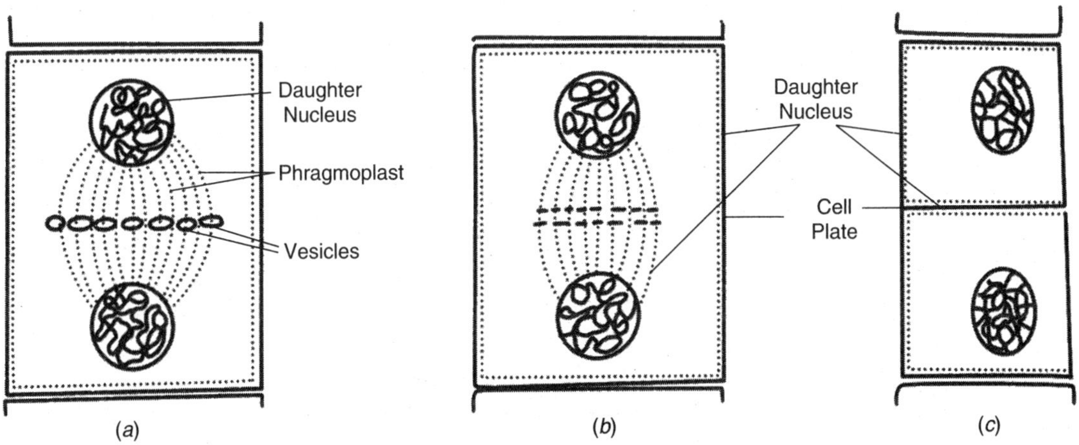

Fig. 6.7. Cytokinesis by cell plate method in plant cells.

When nuclear division takes place without cytoplasmic division it results in the formation of large number of nuclei in a cell. This condition is called **syncytium.**

Differences between Animal Cytokinesis and Plant Cytokinesis

Animal Cytokinesis	*Plant Cytokinesis*
1. It occurs by cleavage.	1. It commonly occurs by cell plate method.
2. Cleavage begins at the periphery and proceeds inward.	2. Cell plate appears at the centre and extends outward.
3. The spindle begins to degenerate soon after the anaphase. It is replaced by midbody.	3. The spindle usually persists during cytokinesis. The persistent spindle apparatus rather grow in size and is called phragmoplast.
4. Cleavage starts by the contraction of peripheral ring of microfilaments.	4. Vesicles derived from Golgi apparatus fuse to form cell plate and new cell membranes.
5. New cell membrane is usually derived from endoplasmic reticulum.	5. The new cell membrane is derived from the vesicles of Golgi apparatus.

Significance of Mitosis

- **Growth.** New somatic cells are formed by mitosis. Therefore, mitosis helps in growth and development of multicellular organisms.
- **Maintenance of Size.** Mitosis helps in maintaining a proper cell size by dividing an over grown somatic cell. Thus, it maintains the surface/volume ratio of the cell.
- **Maintenance of Chromosome Number.** Mitosis ensures equal number of chromosomes in all the cells of an individual. This helps in proper co-ordination among different cells.
- **Reproduction.** Mitosis is a method of multiplication of unicellular organisms.
- **Healing and Regeneration.** Mitosis produces new cells for the healing of wounds and for regeneration.
- **Nucleoplasmic Index.** The growth of multicellular organisms is due to mitosis. A cell cannot grow in size to a large extent without disturbing the ratio between the nucleus and the cytoplasm. After a particular size has been reached, the cell divides to restore the nucleo-plasmic index. Thus, growth takes place by the increase in the number of cells, rather than by increase in the size of the cells. An idea of growth due to mitosis can be obtained from the fact that the human body contains about 6×10^{12} cells, all ultimately derived by the division of a single zygote cell.
- **Repair.** Repair of the body takes place because of the addition of cells by mitosis. The cells of the upper layer of the epidermis, cells of the lining of the gut, and RBCs are constantly being replaced. It is estimated that in the human body about 5×10^9 cells are lost and replaced daily.
- **Opportunity for Differentiation Due to Multicellular Condition.** This is not possible in a single-celled organism.
- **Evidence of Basic Relationship.** The details of mitosis are similar in the majority of organisms, showing their basic similarity and relationships.

Differences between Plant and Animal Mitosis

Plant Mitosis	*Animal Mitosis*
1. It generally occurs in meristematic regions of the plants.	1. It occurs at several places in the animals body.
2. The cell does not change shapes prior to division.	2. The cell often becomes spherical prior to division.
3. The spindle poles lack centriole.	3. Centrioles are present at spindle poles.
4. The spindle is anastral.	4. The spindle is amphiastral *i.e.,* one aster at each pole.
5. The equatorial region of the spindle forms phragmoplast.	5. The equatorial region of the spindle forms mid body.
6. Cytokinesis generally occurs by cells plate method.	6. Cytokinesis occurs by cleavage.
7. Cell plate or middle lamella cements the daughter cells.	7. Cleavage creates an intercellular space between the daughter cells.
8. Plant mitosis is generally controlled by hormone cytokinins.	8. Animal mitosis is controlled by a number of mitogens like lymphokines, epidermal growth factor, platelet derived growth factor etc.

MEIOSIS

Meiosis (Gk. *meo*—to lessen) *is a division which occurs in a diploid cell (or nucleus) and gives rise to four haploid cells or nuclei, each having half the number of chromosomes as compared to the parent cell.* The division was first studied by **Beneden** (1887), **Strasburger** (1888), **Sutton** (1900), **Winiwater** (1900) in different organisms. The term 'meiosis' was first introduced by **Former** and **Moore** (1905).

In meiosis, the nucleus divides twice but the chromosomes are replicated only once. Thus, meiosis reduces the chromosome number as well as the amount of DNA per cell to half. In diploid organisms like higher plants and animals, meiosis occurs during the formation of spores or gametes, whereas in haploid organisms, it occurs during the germination of zygote. The cells in which meiosis takes place are called **meiocytes.** In animals, meiocytes are of two types, spermatocytes and oocytes. In higher plants, meiocytes are differentiated into microsporocytes and megasporocytes.

The essential stages that take place during meiosis are :

1. Two successive divisions without any DNA replication occurring between them.
2. Formation of chiasmata and crossing over.
3. Segregation of homologous chromosomes.
4. Separation of sister chromatids.

Types of Meiosis

Meiosis is essentially similar in all plants and animals. However, depending upon the stage, when meiosis occurs, the latter is of three types—gametic, zygotic and sporic (Fig. 6.8).

1. **Gametic or Terminal Meiosis.** In most of animals and some lower plants meiosis takes place during the formation of gametes (gametogenesis). Such a meiosis is described as **gametic** or **terminal meiosis.** When two gametes fuse in fertilization, the diploid zygote is formed. Gametic meiosis results in **diplontic life cycle.**

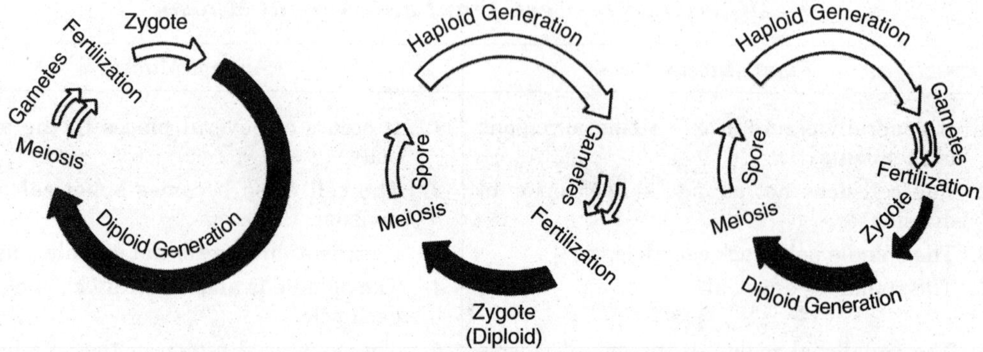

Fig. 6.8. Types of Meiosis.

2. **Zygotic or Initial Meiosis.** In some lower plants (such as fungi, certain protists and some algae) meiosis takes place in the zygote and the resulting organisms are haploid. It is called **zygotic** or **initial meiosis.** Organisms having zygotic meiosis have **haplontic life cycle.**

3. **Sporic or Intermediate Meiosis.** In most of the plants (such as pteridophytes, gymnosperms and angiosperms) meiosis occurs at the time of **sporogenesis** (formation of spores or microspores and megaspores). It is called **sporic** or **intermediate meiosis.** Spore gives rise to a new gametophytic phase in the life cycle. The gametophyte produces gametes. The life cycle of plants is said to be **diplohaplontic** because of the presence of diploid and haploid multicellular phases.

The cells having two genomes are called diploid cells, while those having single genome are called haploid cells. A genome is a complete complement or set of chromosomes where each kind is represented by a single chromosome.

Interphase

Prior to meiosis, there is an interphase. It is similar to the interphase of mitosis except that S-phase is prolonged. A distinct G_2 phase is either short or altogether absent. Therefore, the meiosis division takes place just after DNA and chromosome replication is completed. Like mitosis, meiosis can be divided into two steps—karyokinesis and cytokinesis.

Karyokinesis

Meiosis involves two successive nuclear divi-sions, which occur one after the other (Fig. 6.9). These two divisions are called **meiosis I** and **meiosis II.** During meiosis I, the chromosomes become genetically different due to crossing over and their number is reduced to half. Therefore, meiosis I is also called **heterotypic** or **reduction division.** During meiosis II, the sister chromatids are separated and the chromosome number remains the same as produced after the end of the first division. Therefore, meiosis II is also called **homotypic** or **equational division.**

Meiosis I

Like mitosis the first meiotic division is studied under four phases. These are prophase I, metaphase I, anaphase I and telophase I.

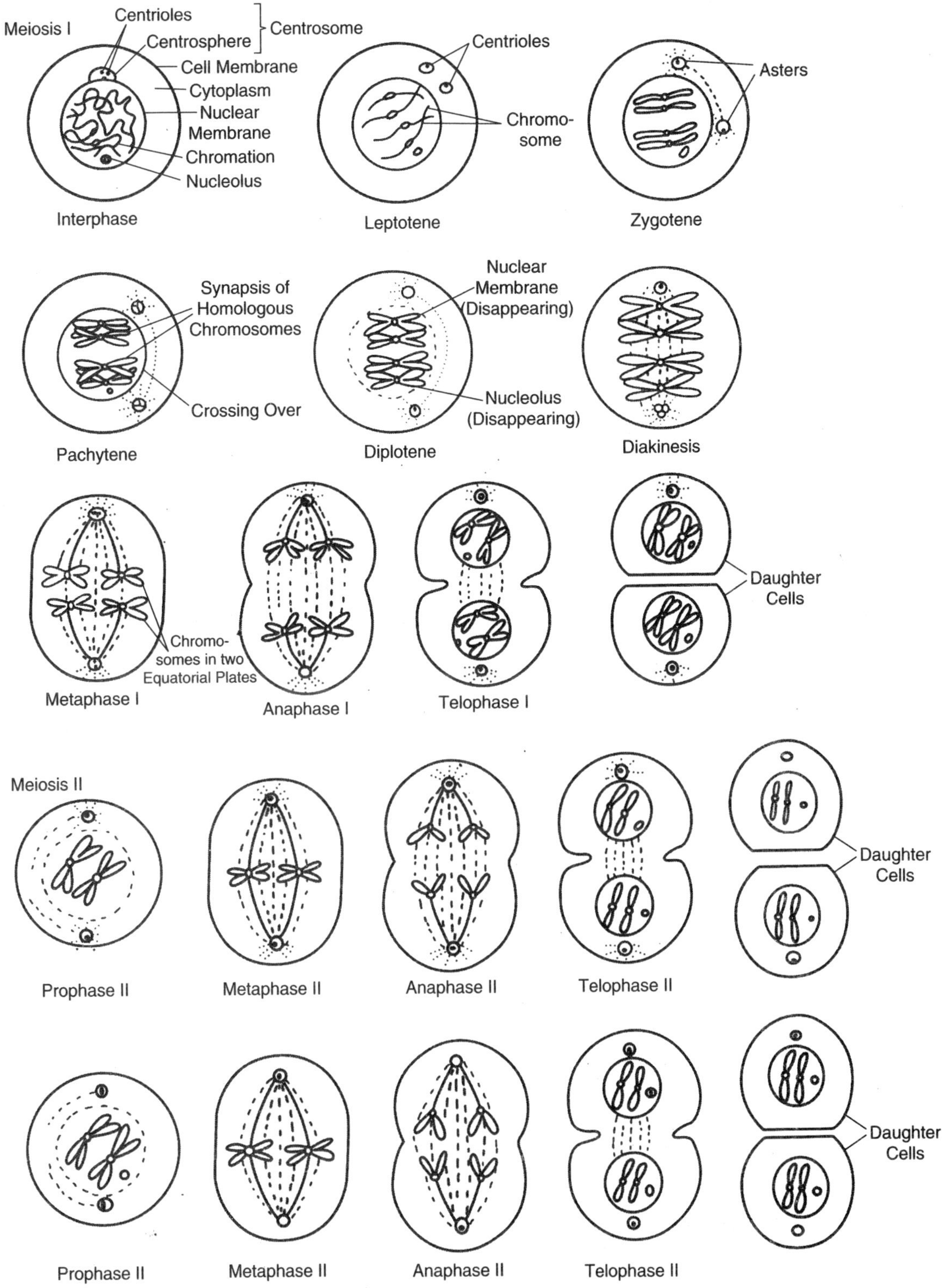

Fig. 6.9. Different stages of Meiosis I in animal cell.

Prophase I

It is more complicated and prolonged as compared to the similar stage of mitosis. It may be divided into the five sub-phases—leptotene, zygotene, pachytene, diplotene and diakinesis.

1. ***Leptotene or leptonema*** (Gk. *leptos*—slender ; *tainia*—band ; *nema*—thread). It is the first stage of meiotic division following the interphase. The chromatin fibres of interphase nucleus shorten and elongated, chromosomes become clear. Each chromosome is attached at both of its ends to the nuclear envelope *via* a specialised structure called **attachment plate.** The chromosomes have already replicated during the interphase, but the chromatids are not distinguishable due to the presence of nucleoprotein core between them. Thus each chromosome appears to be longitudinally single. The chromosomes bear a series of dense granules called **chromomeres.** Chromomeres are often believed to represent genes.

In many animal cells, the ends of the chromosomes may converge towards the side having replicated centrosomes or centriole pairs. This shows a peculiar arrangement called **bouquet stage.** One of the two centriole pairs begin to move to the opposite side. Both the centriole pairs or centrosomes develop astral rays from the pericentriolar satellites. Each centriole pair and its astral rays together constitute aster. There are no centriole in plant cells and no asters are formed.

There are two sets of chromosomes in a cell undergoing meiosis, *i.e.,* the chromosome number is diploid. The chromosomes of one set are contributed by the father parent and are called **paternal chromosomes,** while those of the other set are contributed by the mother parent and are called **maternal chromosomes.** For each paternal chromosomes, there is always present a corresponding maternal chromosome, similar in size, shape and nature of inherited characters. These are called **homologous chromosomes.**

2. ***Zygotene or Zygonema*** (Gk. *zygon*—yoke or tied ; *tainia*—band). The homologous chromosomes come to lie side by side and get attached laterally due to the development of nucleoprotein between them. Pairing is such that the genes of the same character present on the two chromosomes come to lie exactly opposite. This pairing of homologous chromosomes is called **synapsis** or **syndesis.**

It produces a complex, known as **synaptinemal complex.** It is believed to stabilize the paired condition of chromosomes till crossing over is completed. On account of synapsis, chromosomes form pairs or **bivalents.** The number of bivalents is half the number of the total chromosomes.

3. ***Pachytene or Pachynema*** (Gk. *pachys*—thick ; *tainia*—band). The paired chromosomes or bivalent become short and thick. The chromatids of each paired chromosome slightly separate and become visible. Thus, each bivalent or chromosome pair is made up of actually four chromatids, two of each chromosome. Each group of four chromatids is referred as **tetrad.** The two chromatids of the same chromosome are called **sister chromatids,** and those belonging to two different chromosomes of a homologous pair are termed as **non-sister chromatids.**

During pachytene, there occurs exchange of the corresponding segments of non-sister chromatids of homologous chromosomes. This process is called **crossing over.** Crossing over takes place by breakage and reunion of chromatid segments. Breakage, called **micking,** is assisted by an enzyme **endonuclease** and reunion, termed **annealing,** is aided by an enzyme **R-protein** (= **recombinase**). After crossing over, the two chromatids of a chromosome become dissimilar (Fig. 6.10).

Crossing over is a common event during meiosis. It introduces new combination of genes (recom-binations) and results in variations.

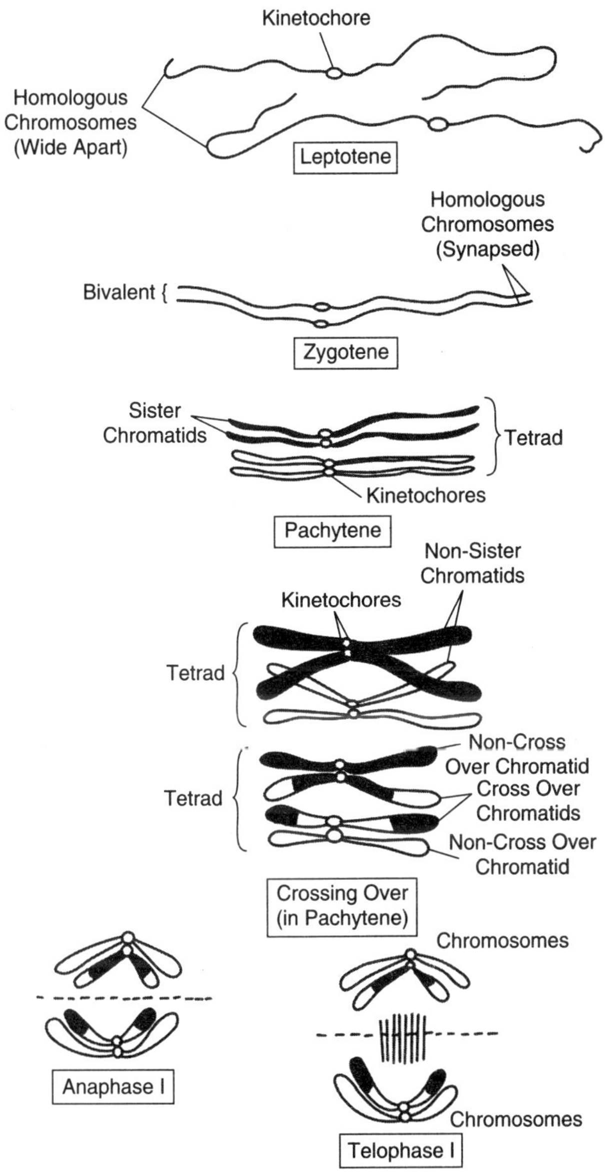

Fig. 6.10. Behaviour of chromosomes in meiosis-I. including crossing over.

4. ***Diplotene or Diplonema*** (Gk. *diplos*—double ; *tainia*—band). The nucleoprotein fusion complex of the synapsed chromosomes dissolves. The homologous chromosomes start separating except in the region of crossing over. It makes chromatids more distinct and the tetrads very clear. The points of attachment between the homologous chromosomes after the partial dissolution of nucleoprotein complex are called **chaismeta** (Sig. Chiasma). The chaismata mark the sites of crossing over. Chiasmata may be **terminal** or **interstitial.** Depending upon their position the homologous chromosomes appear cross like ring or chain like. However, they are not permanent structures. They slide sideways or even disappear at places.

5. ***Diakinesis*** (Gk. *dia*—through ; *kinesis*—movement). The chromosomes condense further to acquire characteristic size and form. The chiasmata disappear by sliding towards the ends of the chromosomes. This phenomenon is called **terminalisation.** The nucleolus/ nucleoli disappear. Nuclear envelope breaks down into vesicles.

In oocytes, diplotene can last for months or years, since it is at this stage that the chromosomes decondense and engage in RNA synthesis.

Metaphase I

A bipolar spindle apparatus appears in the region of nucleus. The spindle fibres are formed of microtubules. The fibres converge towards the two ends called poles. In animal cells, the poles are formed by asters. The bivalents arrange themselves on the equator of the bipolar spindle. Since, there are two centromeres in each bivalent, the centromeres of all the bivalent produce a double metaphasic plate. The distribution of the bivalent is at random so that the individual paternal and maternal chromosomes can face either of the two poles of the spindle. Each chromosome gets attached to the spindle pole of its side by means of a **chromosome fibre** or **tractile fibril** which arises in the region of the centromere. The fibres of homologous chromosomes are always in the opposite direction.

Anaphase I

The two chromosomes of each bivalent separate from one another and pass to the opposite poles of the spindle. This process is called **disjunction** (*dis*-seperate, *junction*-union). The separated chromosomes or **univalents** are also called **dyads** because each of them consists of two chromatids which remain joined with each other at the centromere. In contrast to the single stranded nature of chromosomes in anaphase of mitosis, the chromosomes of anaphase I are double stranded.

At the end of anaphase I, two groups of chromosomes are produced, with each group having half the number (haploid) of chromosomes present in the original cell.

Telophase I

The chromosomes elongates but usually do not form chromatin reticulum. A nucleolus is formed by the satellite chromosome. It is followed by the appearance of nucleoplasm and nuclear envelope. The spindle and the astral rays gradually disappear.

Cytokinesis may or may not follow meiosis I. In some cases telophase I is completely omitted and anaphase chromosomes reaching each pole may progress directly to prophase II.

Significance of Meiosis I

1. It separates the homologous chromosomes to reduce the chromosome number to haploid stage. It is essential for sexual reproduction.

2. It introduces variation by forming new combination of genes and random assortment of paternal and maternal chromosomes.

3. It may at times cause chromosomal mutation by abnormal disjunction.

Meiosis II

In this division, the two chromatids of each chromosome separate from each other and go to separate daughter cells. As a result, the number of chromosomes remains the same as produced by meiosis I. Therefore, meiosis II is also called **homotypic** or **equational division.** Though meiosis II resemble mitosis, it is not exactly mitosis because :

(*i*) It does not follow interphase.

(*ii*) The two chromatids of a chromosome are often dissimilar so that the daughter cells formed after meiosis II are neither similar to each other nor similar to the parent cell.

The main function of homotypic division or Meiosis II is to separate the chromatids of univalent chromosomes which become dissimilar due to crossing over.

An interphase is absent. The state of nucleus between meiosis I and meiosis II having elongated chromosomes is called **interkinesis.** In animal cells, centrosomes or centriole pairs undergo replication during interkinesis or early prophase II. However, replication of DNA is absent.

Meiosis II takes place simultaneously in the two nuclei formed after meiosis I. Meiosis II is divisible into prophase II, mataphase II, anaphase II and telophase II.

Prophase II

This stage of nuclear division is very brief. The dyad chromosomes shorten a little. Nuclear membrane and nucleolus disappear. In animal cells, the centriole pairs develop asters and move to the regions of future spindle poles. In case where telophase I is omitted the prophase II is completely absent.

Metaphase II

Bipolar fibrous spindles appear in the region of dividing nuclei. The chromosomes take up position at the equator of the spindle, forming a metaphase plate. Each chromosome gets connected with both the spindle poles by means of chromosome fibres or tractile fibrils that develop from both the surfaces of its centromere.

Anaphase II

The two chromatids of each chromosome separate completely. They are called **daughter** or **new chromosomes.** The daughter chromosomes move towards the poles along the path of their tractile fibrils. At the end of anaphase II, four groups of chromosomes are produced, each group having haploid number.

Telophase II

The four groups of chromosomes are organised into daughter nuclei. The chromosomes uncoiled (stretch) and form chromatin reticulum. The nuclear membrane and the nucleoli are reconstituted. The spindle fibres degenerate.

Cytokinesis

The division of cytoplasm occurs by furrowing or cleavage in animal cells and by cell plate formation in plant cells. In both cases cytokinesis may be of two types—**successive** and **simultaneous.** In successive type, cytokinesis occurs after each nuclear division. First the diploid parent cell divides by heterotypic division into two haploid cells, which then produce four haploid cells by homotypic division. The four daughter cells may form a linear or isolateral tetrad. In simultaneous type, cytokinesis occurs only when both the nuclear divisions are completed. It gives rise to four haploid cells arranged tetrahedraly, collectively called tetrahedral tetrad.

Significance of Meiosis

- **Formation of Gametes.** Meiosis produces gametes that are essential for sexual reproduction.
- **Maintenance of Chromosome Number.** Meiosis maintains the chromosome number constant in sexually reproducing organisms. It is essential since the chromosome number is doubled after fertilization.
- **Introduction of Variation.** Meiosis introduces variation in the organisms due to crossing-over and random distribution of paternal and maternal chromosomes.
- **Mutations.** Chromosomal and genomatic mutations can take place by irregularities of meiotic divisions. Some of the mutations may be advantageous to the organism and are perpetuated by natural selection.
- **Genetic Information.** It switches on genetic information for the development of gametes or gametophyte and switches off the sporophytic information.
- **Evidence of Basic Relationship of Organisms.** Details of meiosis are essentially similar in the majority of organisms showing their basic similarities and relationship.

Differences between Mitosis and Meiosis

Mitosis	*Meiosis*
1. Mitosis takes place in the somatic cells.	1. Meiosis occurs either in the reproductive cells at the time of formation of gametes or spores or at the time of zygote germination.
2. It may occur both in haploid and diploid cells.	2. It occurs only in diploid cells.
3. After one mitosis, daughter cells may undergo, further mitosis.	3. Meiosis cannot be repeated by the daughter cells.
4. Mitosis is a single division which produces two cells.	4. Meiosis is a double division, which gives rise to four cells.
5. Subsequent mitosis divisions are similar to the earlier ones.	5. The two divisions of meiosis are not similar. The first one is heterotypic or reductional, while the second one is homotypic or equational.
6. Each mitosis division is preceded by an interphase.	6. The second meiotic division is not preceded by an interphase.
7. The chromosomes replicate in the interphase before every division.	7. The chromosomes replicate only once, prior to meiosis.
8. The number of chromosomes remains the same after mitosis.	8. The number of chromosomes is reduced to half after meiosis.
9. The daughter nuclei or cells formed after mitosis are exactly similar to the parent one.	9. The daughter nuclei or cells formed after meiosis are neither similar to the parent one nor to one another.
Prophase	
10. Prophase is relatively short and without sub stages.	10. Prophase I is of long duration and is divisible into five substages, Prophase II is however, brief and simple.
11. Prophase chromosomes appear double stranded from the very beginning.	11. Prophase I chromosomes do not look double stranded in the beginning. The two chromatids of each chromosome become clear during diplotene of prophase I.
12. Pairing of chromosome does not occur in mitosis.	12. Pairing or synapsis of homologous chromosomes takes place during zygote of prophase I.

13. A synaptinemal complex is absent.	13. Synapsed homologous chromosomes develop a synaptinemal complex.
14. Crossing over does not occur hence chiasmata are absent.	14. Crossing over usually takes place during pachytene and chiasmata are observed during diplotene and diakinesis of porphase I.
Metaphase	
15. Single metaphasic plate is formed by all the chromosomes.	15. A double metaphase plate is formed by the chromosomes in metaphase I but only one in metaphase II.
16. The centromere of a chromosome connected with both the spindle poles.	16. The centromere of a chromosome is connected to one spindle pole in metaphase I but both in metaphse II.
17. Two chromatids of a chromosome are genetically similar.	17. The two chromatids of a chromosome are often genetically different due to crossing over.
Anaphase	
18. Anaphase involves separation of chromatids of each chromosome.	18. Anaphase I involves separation of homologous chromosomes, the chromatids are separated in anaphase II.
19. Similar chromosomes move towards the opposite poles in anaphase.	19. Dissimilar chromosomes move towards the opposite poles.
Telophase	
20. Telophase is longer and produce interphase nuclei.	20. Telophase I is shorter and nuclei never enter the interphase.
Cytokinesis	
21. Cytokinesis follows every mitosis.	21. Cytokinesis may occur at the end of each division or simultaneously after meiosis II.

CONTROL OF CELL CYCLE

The various phases of cell cycle are controlled by proteins called **cyclins** and **cyclin dependent protein kinases (CDKs).** The kineses are the enzymes that phosphorylate proteins using ATP. The CDKs bring about transition from G_1 to S phase and S to M phase. In fact, there are two regulatory mechanisms, called **check points,** which take decision about cell division (Fig. 6.11). The **first check point,** called G_1 cyclin (CG_1 or G_1/S) lies in between G_1 and S. In G_1, CDK becomes active by G_1 cyclin and ATP at its activation site. It causes transition of G_1 to S phase. G_1 cyclin is destroyed at the end of S phase and CDK becomes inactive. Cell enters G_2 phase. **Second check point** lies between G_2 and M. In G_2 inactive CDK binds to second check point, mitotic cyclin (C_M and G_2/M). It gets phosphosylated at its both activation and inhibatory sites but still remains inactive. On removal of phosphate (PO_4) from the inhibitory site, it gets active and cause transition from G_2 to M phase. At the end of M phase, cyclin C_M is degraded and phosphate (PO_4) at activation site is removed. The cell enters G_1 phase again.

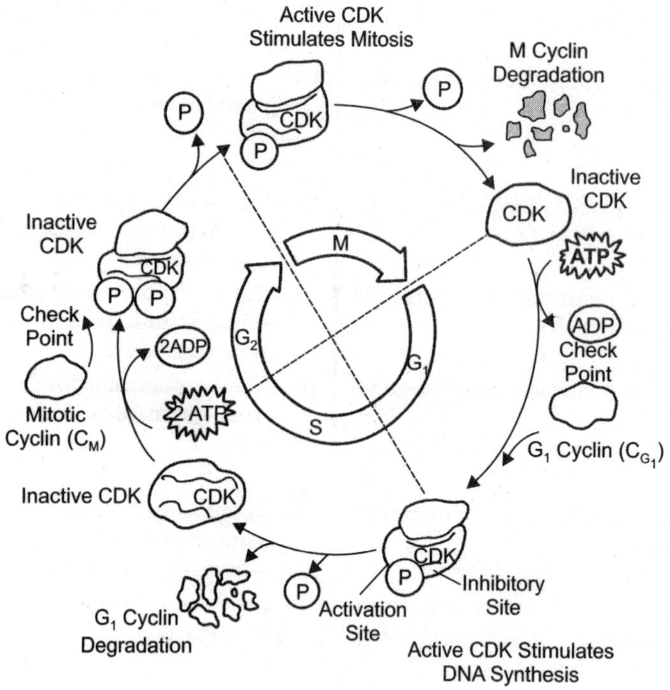

Fig. 6.11. Diagram of the cell cycle showing regulation of the cell cycle by Cyclin Dependent Protein Kinase (CDK).

REVIEW EXERCISES

MULTIPLE CHOICE TYPE QUESTIONS

1. 'Nothing lives forever yet continues' explain the role of
(*a*) Adaptation in nature (*b*) Decomposition in nature
(*c*) Reproduction in nature (*d*) Nutrition in nature.

2. Reproduction is essential for
(*a*) Producing young ones (*b*) Continuity of a species
(*c*) Adaptations (*d*) Variations.

3. DNA replication occurs during
(*a*) G_1 phase (*b*) G_2 phase
(*c*) S-phase (*d*) Prophase.

4. The term meiosis was coined by
(*a*) Flemmig (*b*) Blackman
(*c*) Former and Moore (*d*) Robertson.

5. Significance of mitosis is
(*a*) Quick divisions (*b*) Increasing cellular mass
(*c*) Occurrence in every tissue
(*d*) Production of cells, genetically similar to the parent cell.

6. Stage of mitosis in which the chromosomes move towards the poles is
(*a*) Prophase (*b*) Metaphase
(*c*) Anaphase (*d*) Telophase.

7. Chismata are first observed in
   (*a*) Zygotene (*b*) Leptotene
   (*c*) Diplotene (*d*) Pachytene.
8. The number of mitotic divisions required to produce 128 cells from a single cell is
   (*a*) 7 (*b*) 14
   (*c*) 16 (*d*) 32.
9. If after mitotic division cell activity is restricted to G_1-phase of the cell cycle, then the condition is known as
   (*a*) G_2-phase (*b*) G_0-phase
   (*c*) S-phase (*d*) M-phase.
10. Identify the incorrect statement about meiosis.
   (*a*) Pairing of homologous chromosomes
   (*b*) Four haploid cells are formed
   (*c*) At the end of meiosis the number of chromosomes are reduced to half
   (*d*) Two cycles of DNA replication occurs.

OBJECTIVE TYPE QUESTIONS

1. Fill in the blanks with suitable word.
   (*i*) Somatic cells multiply by .... division.
   (*ii*) Mitosis results in the formation of nuclei having .... number of chromosomes.
   (*iii*) The second division of meiosis may be described as .... division.
   (*iv*) The region of attachment of chromosomes to spindle fibres is called ....
   (*v*) In meiosis, haploid condition is reached by .... stage.
2. Supply a specific scientific term for each of the following.
   (*i*) The period between two successive mitotic divisions.
   (*ii*) Process of cell division by which chromosome number is halved.
   (*iii*) Point at which two sister chromatids are held together.
   (*iv*) Phase in the cell cycle when proteins and RNA is synthesised.
   (*v*) The nuclear division in which chromosome number remains equal to the parent cell.
3. Match the items listed in Column I with suitable words from Column II.

Column I	Column II
(*i*) Karyokinesis	(*a*) Diplotene
(*ii*) Cell plate	(*b*) Pachytene
(*iii*) Cytokinesis	(*c*) Nuclear division
(*iv*) Crossing over	(*d*) Plant cells
(*v*) Chiasmata	(*e*) Cytoplasmic division

4. Assign the following statements to the (*i*) Prophase (*ii*) Metaphase (*iii*) Anaphase and (*iv*) Telophase of mitotic division.
   (*i*) The nuclear membranes reappear.
   (*ii*) Chromosomes are thickest and shortest.
   (*iii*) Chromosomes begin to coil.
   (*iv*) Centromere divides into two.
   (*v*) Followed by cytokinesis.
   (*vi*) Each chromosome consists of two chromatids.

TEST QUESTIONS

1. What is reproduction ? What is its significance in the living world ?
2. Differentiate between asexual reproduction and sexual reproduction.
3. Why is mitosis called equational division ?
4. What is G_0 (quiescent phase) of cell cycle ?
5. Describe the events taking place during interphase.
6. Give the sequence of events occurring during prophase of mitosis.
7. What is the difference between Cytokinesis in an animal cell and a plant cell.
8. Describe the changes occurring in the nucleus during prophase I of mitosis.
9. Give a short account of various stages of mitosis.
10. List the differences between mitosis and meiosis.

UNIT–4

GENETIC ENGINEERING

7 Introduction to Genetic Engineering

Genetic engineering also referred as **'recombinant DNA technology'** or **'gene splicing'**, is a kind of biotechnology involving manipulation of DNA. If deals with the isolation of useful genes from a variety of sources and the formation of new combinations of DNA (called **recombinant DNA** or **rDNA**) for repair and improvement of a genotype. Genetic engineering alters the chemistry of genetic material (DNA and RNA) to introduce these into host organisms to bring about a change in the phenotype of the host organism. Thus, genetic engineering may be defined *'as a technique for artificial and deliberately modifying DNA (gene) to suit human needs'* **Paul Berg** (1972), was able to introduce a gene of SV-40 into a bacterium with the help of lambda phage (λ-phase). Paul Berg is often considered **'Father of Genetic Engineering'.** He was awarded Nobel Prize in 1980.

Conceptual Principles of Genetic Engineering. You must be aware that sexual reproduction is more advantageous than asexual reproduction. Sexual reproduction provides opportunities for variation and formulation of unique combinations of genetic make up. Some of the variation may be beneficial to the organism as well as the population. Asexual reproduction preserve the genetic traits, while sexual reproduction permits variation. Traditional methods of hybridization used in plant and animal breeding, very often lead to inclusion and multiplication of both undesirable and desirable genes in the hybrids. However, the techniques of genetic engineering involving creation of **recombinant DNA,** use of **gene cloning** and **gene transfer,** overcome this limitation, and permits us to isolate and introduce only one or a set of desirable genes without introducing undesirable genes into the target organism.

When a piece of DNA is introduced into an alien (foreign) organism, it would not be able to multiply itself in the progeny cells of the organism. But, when it gets integrated into the genome of the recipient, it may multiply and be inherited along with the host DNA. It is because the alien piece of DNA has become part of a chromosome, which has the ability to replicate. There is a specific DNA sequence in the chromosome called **origin of replication,** which is responsible for initiating replication. Therefore, for the replication of any alien piece of DNA, it must joint a chromosome which has an origin of replication (ori), so that this alien piece of DNA can replicate and multiply itself in the host organism. This is referred to as **cloning** or making identical copies of any template DNA.

Construction of Artificial Recombinant DNA Molecule. The construction of the first recombinant DNA emerged from the possibility of linking a gene encoding antibiotic resistance with a native **plasmid** of *Salmonella typhimurium*. Plasmid is self replicating extra chromosomal circular DNA present in some bacteria. **Stanley Cohen** (Stanford University, USA) and **Herbert Boyer** (University of California, USA) constructed first recombinant DNA in 1972, by isolating the antibiotic resistance gene by cutting out a piece of DNA from a plasmid which was responsible for giving antibiotic resistance. The cutting of DNA at a specific locations became possible with the discovery of **restriction enzymes** popularly called **'molecular scissors'.** The cut piece of DNA was then linked with the plasmid DNA. These plasmid DNA act as **vectors** to transfer the piece of DNA attached to it. As you know that female *Anopheles* mosquito acts as an insect vector to transfer the malarial parasite into human body. In the same way, a plasmid can be used as vector to transfer an alien piece of DNA into the host organism. The linking of antibiotic resistance gene with the plasmid vector is done with the help of enzyme **DNA ligase.** DNA ligase acts on cut DNA molecules and join their ends. This makes a new combination of circular autonomously replicating (self replicating) DNA, created *in vitro*. It is known as **recombinant DNA.** When this recombinant DNA is transferred into *Escherichia coli,* a bacterium closely related to *Salmonella,* it could replicate using the new host's DNA polymerase enzyme and make multiple copies. The making of multiple copies of antibiotic resistance gene in *E. coli* was termed as **cloning** of antibiotic resistance gene in *E. coli.*

TOOLS OF RECOMBINANT DNA TECHNOLOGY

The Key tools used in genetic engineering or Recombinant DNA technology include restriction enzymes, polymerase enzymes, ligases, vectors and the host organism.

Restriction Enzymes (The Molecular Scissors)

Restriction enzymes belong to a larger class of enzymes called **nucleases.** These are of two kinds: **exonucleases** and **endonucleases**, Exonucleases remove nucleotides from the ends of the DNA, whereas, endonucleases make cuts at specific positions within the DNA. Restriction enzymes are present in many bacteria where they function as a part of their defence mechanism called the **restriction modification system (Wemer Arber,** 1965). This system comprises of two types of enzymes: (*i*) **Restriction enzymes.** They identify the introduced foreign DNA and cut it into pieces. They restrict the propagation of foreign DNA (*e.g.,* bacteriophages) in a bacterium and are therefore, also called **restriction endonucleases** (*ii*) **Modification enzyme.** It adds methyl group to one or two bases usually within the sequence recognised by the restriction enzyme. The DNA having modified sequences cannot be recognised and cut by restriction enzymes. This is how a bacterium protects its own chromosomal DNA from cleavage by these restriction enzymes.

There are three classes of restriction endonucleases. Type I, Type II and Type III, differing slightly in their mode of action. Out of these only Type II restriction enzymes are used in recombinant DNA technology, because they can be used *in vitro* to recognise and cut within specific DNA sequence typically consisting of 4 to 8 nucleotides. The first restriction endonuclease was *Hin*d II. Its functioning depended on a specific DNA nucleotide sequence. It was isolated and characterised five year later. *Hin*d II always cut DNA molecules at a particular point by recognising a specific sequence of six base pairs. This specific base sequence is known as the **recognition sequence** for *Hind* II. Besides *Hind* II, presently, more than 900 restriction enzymes are known, which have been isolated from over 230 strains of bacteria, each of which recognise different recognition sequences.

Nomenclature of Restriction Enzymes. The convention for naming restriction enzymes is based on the name of bacterium from which they have been isolated. The first letter of the name comes from the genus (in italics), and second two letters (also in italics) come from the species of the bacterium, next is the strain of the organism, last in a Roman numeral signifying the order in which the enzyme was isolated from that strain of the bacteria. *e.g., ECO*RI comes from *Escherichia coli* RY 13. In *ECO*RI, the letter *'E'* stands for *Escherichia co for coli*, 'R' for the strain of the bacterium and it was first endonuclease (I) from this bacterium. **W. Arber, H. Smith** and **D. Nathans** received Nobel prize in 1978, for the discovery of this enzyme. Some enzymes commonly used in recombinant DNA technology their source and their recognition base sequence (indicated with arrow) are given in Table 7.1.

Table 7.1

Restriction Enzyme	*Source (Organism and Strain)*	*Sequence Recognised*
1. *Ava* I	*Anabaena variabilis* ATCC 27892	↓ C Py C G Pu G * G Pu G C Py C ↑
2. *Alu* I	*Arthrobacter luteus*	↓ A G C T T C G A ↑
3. *Bam* HI	*Bacillus amyloliquefaciens* H	↓ G G A T C C C C T A G G ↑
4. *Eco* RI	*Escherichia coli* RY 13	↓ G A A T T C C T T A A G ↑
5. *Eco* R II	*E. coli* R 245	↓ C C A (T) G G * * G G T (A) C C ↑
6. *Hind* II	*Haemophilus influenzae* Rd	↓ G T Py Pu A C C A Pu Py T G ↑
7. *Hind* III	*H. influenzae* Rd	↓ A A G C T T T T C G A A ↑

Enzyme	Source	Recognition sequence
8. *Hin* f I	*H. influenzae* Rf	↓ G A N T C *** C T N A G ↑
9. *Nla* III	*Neisseria lactamica*	↓ C A T G G T A C ↑
10. *Pst* I	*Providencia stuartii*	↓ C T G C A G G A C G T C ↑

* Pu – Any of the two purines (adenine or guanine) ; Py any of the two pyrimidines (thymine or cytosine).

** Base given in parenthesis *e.g.,* (T) may occur in place of preceding base *e.g.,* A.

*** N – Any of the four bases (adenine, guanine, thymine or cytosine).

Action of Restriction Endonuclease. Each restriction endonuclease functions by inspecting the length of a DNA sequence. Once it finds its specific recognition sequence, it will bind to the DNA and cut each of the two strands of the double helix at specific points in their sugar phosphate backbones (Fig. 7.1). Each restriction endonuclease recognises a specific **palindromic nucleotide sequences** in the DNA. The palindrome is a group letters that reads the same forwards and backwards, *e.g.,*

MALAYALAM

The palindrome in DNA is a sequence of base pairs that reads same on the two strands when orientation of reading is kept the same. For example, the following sequences read the same on the two strands in $5' \rightarrow 3'$ direction. This is also true when we read in the $3' \rightarrow 5'$ direction.

5′– G AA TT C – 3′ Palindromic sequence

3′– C TT AA G – 5′

Several restriction enzymes cut strands of DNA a little away from the centre of the palindrome sites, but between the same two bases on the opposite strands. As a result, the DNA fragments produced by the cut of these enzymes have short, single stranded over hangs at each end. Such ends are known as **sticky ends.** These are named so because they form hydrogen bonds with their complementary cut counter parts. This stickiness of the ends helps enzyme DNA ligase in joining DNA fragments having the same kind of sticky ends. When cut by the same restriction enzyme, the resultant DNA fragments have the same kind of 'sticky ends' and these can be joined together end to end using DNA ligase (Fig. 7.2). In genetic engineering, restriction endonucleases are used to form recombinant DNA molecules, composed of DNA from different sources.

Some Other Enzymes Used in Recombinant DNA Technology

Besides restriction enzymes, several other enzymes also play significant role in recombinant DNA technology. Some of them are as follows :

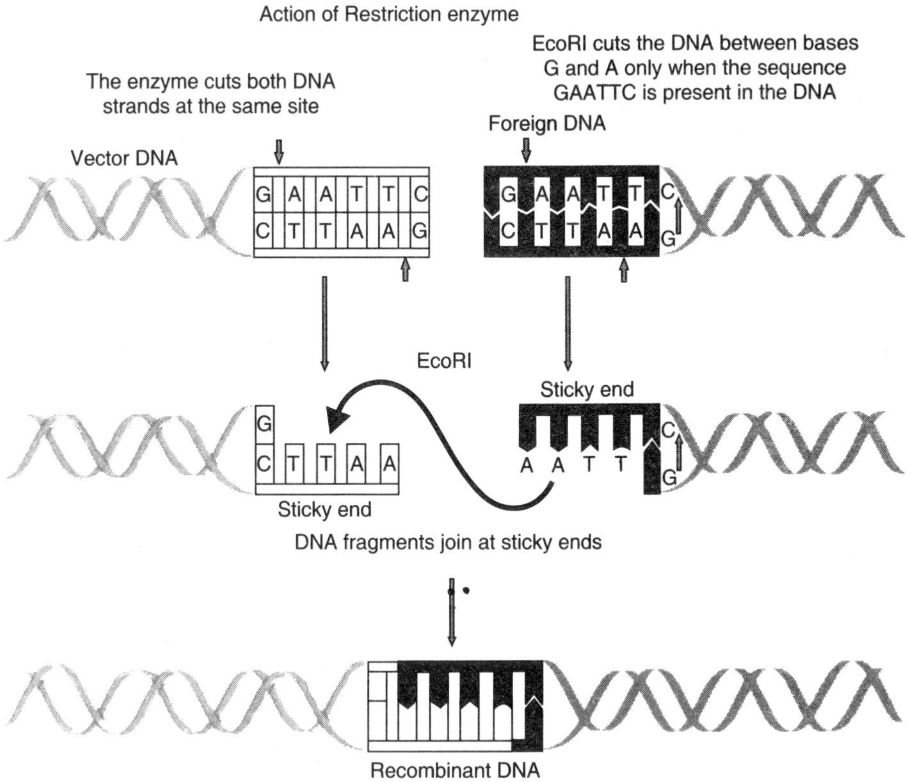

Fig. 7.1. Steps in formation of recombinant DNA by action of restriction endonuclease enzyme-Eco RI.

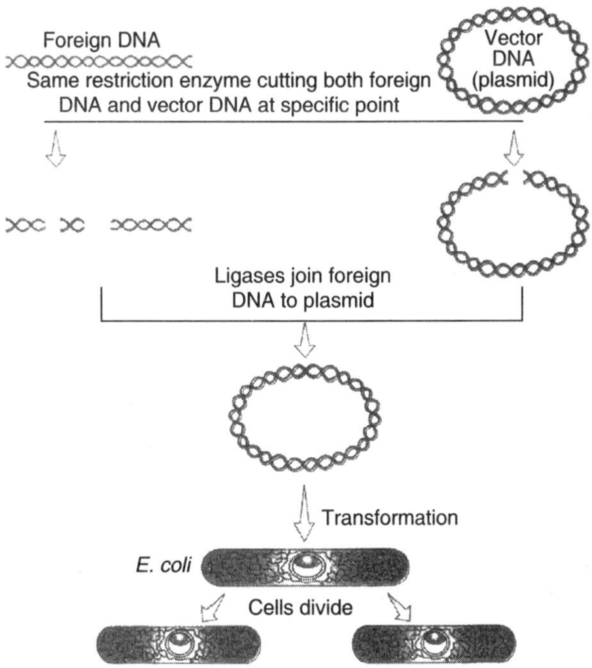

Fig. 7.2. Diagrammatic representation of recombinant DNA technology.

1. DNA Ligase (Molecular glue)

It helps in joining the DNA fragments and is, therefore, commonly called **'molecular glue'.** DNA ligase forms phosphodiester bonds between adjacent nucleotides and covalently links two individual fragments of double stranded DNA. The enzyme requires a phosphate at the 5′ carbon of one nucleotide and a hydroxyl group at the 3′ carbon of the adjacent nucleotide to form the phosphodiester bond between these two nucleotides. In rDNA technology, most commonly used joining enzyme is T_4 DNA ligase, encoded by phage T_4.

2. Alkaline Phosphatase (AP)

The ligation (joining of DNA segments) require the presence of 5′ phosphate group at the DNA site to be ligated. Removal of phosphate group from the 5′ end of a DNA molecule prevents DNA ligation. Alkaline phosphate removes the phosphate group from the 5′ end of a DNA molecule, leaving a free 5′ hydroxyl group. This enzyme is, therefore, used to prevent unwanted self ligation of vector DNA molecules in procedures of rDNA technology. Alkaline phosphate (AP) is isolated from bacteria (called BAP) or calf intestine (called CAP).

3. DNA Polymerase

It help in the synthesis of complementary DNA (cDNA) strand on DNA templates. **A. Kornberg** *et.al.,* (1956), have identified three types of DNA polymerases in *E. coli.* These are—DNA polymerase I (DNA pol I), DNA polymerase II (DNA pol II) and DNA polymerase III (DNA pol III). They have almost similar catalylic activity, however, DNA polymerase III is more active than the other two.

Cloning Vectors (Vehicles for Cloning)

The DNA molecule that can carry foreign DNA segment and replicate inside the host cell is called **vector.** The vectors that may be used in genetic engineering include plasmids, bacteriophages, Yeast Artificial Chromosomes (YACs), Bacterial Artificial Chromosomes (BACs), transposons (jumping genes), cosmids, phagemids, animal and plant viruses and shuttle vectors. However, plasmids and bacteriophages are most commonly used vectors.

1. Plasmid Vectors

Plasmids were discovered by **W. Hays** and **J. Lederberg** (1952) in bacterial cells. They are small extra-chromosomal self-replicating usually circular double stranded DNA molecules that occur naturally in many bacteria and some yeasts. Although plasmids are usually not essential for normal cell growth and division, they often confer some traits on the host organism. *e.g.,* resistance to certain antibiotics or toxins that can be a selective advantage under certain conditions. Some plasmids may have only one or two copies per cell whereas others may have 15–100 copies per cell. Their numbers can go even higher. If we are able to link an alien piece of DNA with plasmid DNA, we can multiply its numbers equal to the copy number of the plasmid. At present plasmids used as vectors are engineered in such a way that they help easy linking of foreign DNA and selection of recombinants from non-recombinants.

One of the easily manipulated plasmid used as vector is **pBR 322.** It is the first artificial cloning vector constructed by **Boliver** and **Rodriguez** in 1977. In the name of plasmid, the letter P signifies plasmid, B and R stand for the initials of the two scientists (Boliver and Rodriguez) who developed it. The plasmid contains origin of replication (*ori*) which permits production of multiple copies per cell. It also has two selectable markers (antibiotic resistance genes) *i.e.,* tetracycline, tet^R and ampicillin, amp^R. It also contain unique recognition sites for 12 restriction enzymes (endonucleases). Two unique sites, *Pst* I and *Pvu* I are present within the amp^R gene and *Bam* HI, *Sal* I are within tet^R gene (Fig. 7.3).

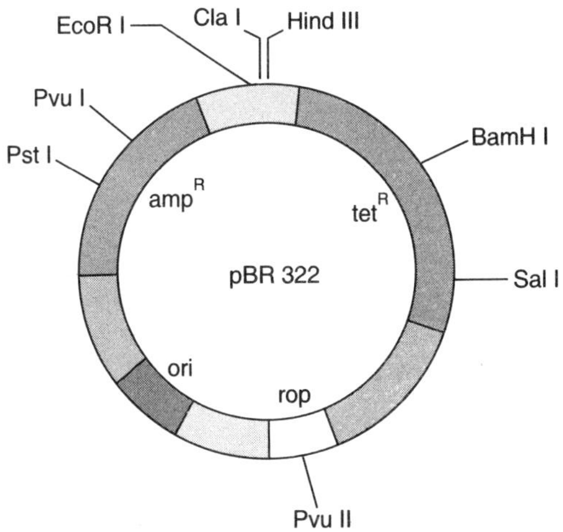

Fig. 7.3. *E. coli* cloning vector pBR 322 showing restriction sites (*Hind* III, *Eco* RI, *Bam*HI, *Sal* I, *Pvu* II, *Pst* I, *Cla* I), ori and antibiotic resistance genes (amp^R and tet^R). Rop codes for the proteins involved in the replication of the plasmid.

The presence of restriction sites within the markers tet^R and amp^R help us to select the cells transformed with the recombinant pBR 322. When enzyme *Pst* I or *Pvu* I is used for the insertion of DNA fragment, the latter is inserted within the gene amp^R. It makes is amp^R non-functional. The bacterial cells containing such a recombinant pBR 322 will not be able to grow in the presence of amplicillin, but will be able to grow in presence of tetracycline. In the same way, when restriction enzyme *Bam* HI or *Sal* I is used, the DNA insert is placed within the gene tet^R making the latter non-functional. The bacterial cells containing such a recombinant pBR 322 will, therefore be able to grow in presence of ampicillin, but not in presence of tetracycline.

2. Bacteriophage Vectors

Bacteriophages (or phages) are viruses that infect bacteria by introducing DNA into the bacterial cells. The injected DNA replicates and expresses in the host bacterial cell resulting in a number of phages which burst out of the cell, reinfect the neighbouring cells (Lytic pathway). This property of bacteriophages to transfer their DNA into specific bacterial cells during the process of bacterial infection gave the scientists the idea that specially designed phage vectors (containing foreign DNA) would prove best tools for gene cloning. Two phages—lambda (λ) and M 13 phage have been extensively modified to be used as cloning vectors.

(*i*) ***Lambda (λ) phage*** has a double stranded DNA having 48,514 bp, in which 12 bases at each end are unpaired but complementary. Such ends are therefore, cohesive (sticky) and are called **Cos sites.** These sites can be used for packaging DNA into phage head. An unique feature of λ genome is that a large fragment in the central region of its genome is not essential for lytic infection of *E. coli*. Therefore, the vector has been designed such that this region can be replaced by a foreign DNA.

(*ii*) ***M 13 phage*** is a filamentous phage which has a single stranded circular DNA of 6407 bp. The M 13 phage DNA, after entry into the bacterial cell, is converted into double stranded molecule called **Replicative Form** (RF), which replicates until there are about 100 copies in the cell. At this stage, the phage genome produces single stranded copies and extrude from the bacterial cell as M 13 particles. The RF can be manipulated like plasmids, while single stranded form is useful for techniques like DNA sequencing and site-directed mutagenesis.

3. Some Other Vectors

Some other vectors which are used in genetic engineering are (*i*) *Yeast Artificial Chromoses* (*YACs*) and *Bacterial Artificial Chromosomes* (*BACs*) are used to clone larger DNA fragments and mapping of large genomes (*e.g.,* HGP) (*ii*) *Cosmids (Cos + plasmid)* are constructed by combining certain features of plasmid and 'Cos' sites of phage lambda and are used to clone DNA fragments upto 45 kb in length. (*iii*) *Phagemids* are composite structures made of bacteriophage and plasmid, and are used for carrying larger DNA sequences and (*iv*) *Animal and plant viruses* are also used to introduce foreign genes into eukaryotic cells in culture.

Features of Cloning Vector

The features that are required in a vector to facilitate cloning of a gene are as follows:

1. **Origin of Replication (Ori).** The sequence in a DNA molecule from where replication starts is called **origin of replication** or **'ori'.** Any piece of DNA when linked to this sequence can be made to replicate within the host cells. This sequence is also responsible for controlling the copy number of the linked DNA. Therefore, when many copies of the target DNA are required, it is cloned in a vector whose origin support high copy number.

2. **Selectable Marker.** The vector also requires a selectable marker. It helps in identifying and eliminating non-transformants by selectively permitting the growth of the transformants. A bacterium which has modified due to the introduction of a piece of foreign DNA is called **transformant.** The process through which a piece of DNA is introduced in a host bacterium is called **transformation.** Normally, the genes encoding resistance to antibiotic such as ampicillin, chloramphenicol, tetracycline or kanamycin etc., are considered useful selectable markers for *E. coli.* The normal *E. coli* cells do not carry resistance against any of these antibiotics.

3. **Cloning Sites (Recognition Sites).** The vector must have very few preferably single recognition sites for the activity of commonly used restriction enzymes in order to link the alien or foreign DNA. The presence of more than one recognition sites within the vector will produce several fragments, which will make the process of gene cloning more complicated. The ligation of foreign DNA is carried out at a restriction site present in one of the two **antibiotic resistance** genes. For example, a foreign DNA can be joined at the *Bam* HI site of tetracycline resistance gene in the vector pBR 322.

Selection of Recombinants. The recombinant vectors can be selected by two methods.

(*i*) *Antibiotic Resistance.* The recombinant plasmids lose tetracycline resistance due to insertion of foreign DNA. It can be selected out from non-recombinants on ampicillin containing medium. The transformants growing on ampicillin containing medium are then transferred on a medium containing tetracycline. The recombinants will grow in ampicillin containing medium but not on that containing tetracycline. However, non-recombinants will be able to grow on the medium containing both the antibiotics. In the present case, one antibiotic resistance gene helps in selecting the transformants but the other antibiotic resistance gene becomes inactivated due to insertion of foreign DNA and helps in selection of recombinants.

(*ii*) *Colour Reaction.* The selection of recombinants due to inactivation of antibiotic resistance gene is a tedious procedure, because it requires simultaneous plating on two plates having different antibiotics. Therefore, alternative selectable marker is developed to differentiate recombinants and non-recombinants on the basis of their ability to produce colour in the presence of a **chromogenic substrate.** In this process, a recombinant DNA is inserted in the coding sequence of an enzyme, **α–galactosidase.** This causes inactivation of the enzyme. This is called **insertional inactivation.** If the plasmid in the bacterium does not have an insert, the presence of a chromogenic substrate gives blue coloured colonies. The presence of insert results into insertional inactivation of the α–galactosidase and the colonies do not produce any colour. Such colonies are marked as recombinant colonies.

4. **Vectors for Cloning Genes in Plants and Animals.** There are procedures of transferring genes into plants and animals from bacteria and viruses. These bacteria and viruses transform eukaryotic cells and force the latter to do as required by the bacteria and viruses. For example, *Agrobacterium tumifaciens,* a pathogen that produces **crown galls** or **plant tumours** in almost all dicot plants (Fig. 7.4). The bacterium contains a large Ti-plasmid (tumour inducing plasmid) and it is able to transfer a piece of DNA known as T-DNA to convert normal cells into a tumour, and direct these tumour cells to produce the chemicals required by the pathogen. Similarly, retroviruses in animals including humans are able to change normal cells into cancerous cells. Presently, the art of delivering genes by pathogens in their eukaryotic hosts is understood in a better way. It enabled scientist to develop techniques to transform the tools of pathogens into useful vectors for delivering genes of interest of humans. The tumour inducing (Ti) plasmid of *Agrobacterium tumifaciens* has now been modified into a cloning vector which is no more pathogenic to the plants but is still able to use the mechanisms to deliver genes of our interest into a variety of plants. Similarly, retroviruses have also been made non-pathogenic and are now used to deliver desirable genes into animal cells. Thus, once a gene or DNA fragment is joined to a suitable vector, it is transferred into a bacterial, plant or animal host, where it undergoes multiplication.

Fig. 7.4. Crown tumour produced by *Agrobacterium tumifaciens.*

STEPS/PROCESSES OF RECOMBINANT DNA TECHNOLOGY

The sequencewise steps involved in recombinant DNA (rDNA) technology includes :

(*i*) isolation of DNA

(*ii*) fragmentation of DNA by restriction endonucleases

(*iii*) isolation of a desired DNA fragment

(*iv*) ligation of the DNA fragment into a vector

(*v*) transferring the recombinant DNA into the host

(*vi*) culturing the host cells in a medium at large scale, and

(*vii*) extraction of the desired product.

1. Isolation of the Genetic Material (DNA)

In majority of organisms, the genetic material is Deoxyribose Nucleic Acid (DNA). The DNA must be in pure form and free from other macromolecules, when it is to be cut with restriction enzyme. Since the DNA is enclosed within the membranes and cell wall, it has to break the cell open to release DNA and other macromolecules like RNA, proteins, polysaccharides and lipids. Therefore, the bacterial cells/plant or animal tissues are treated with enzymes such as **lysozyme** (bacteria), **cellulase** (plant cells), **chitinase** (fungus) to release DNA along with other macromolecules. DNA molecules are interwined with histone proteins, which can be removed by treating it with enzyme **protease.** Associated RNA if any, can be removed by treating with enzyme, **ribonuclease.** The other molecules can be removed by appropriate treatments. The purified DNA, finally precipitates out after the addition of chilled ethanol. The precipitated DNA can be separated out as fine threads by spooling. *i.e.*, winding of fine threads of DNA on a reel (Fig. 7.5).

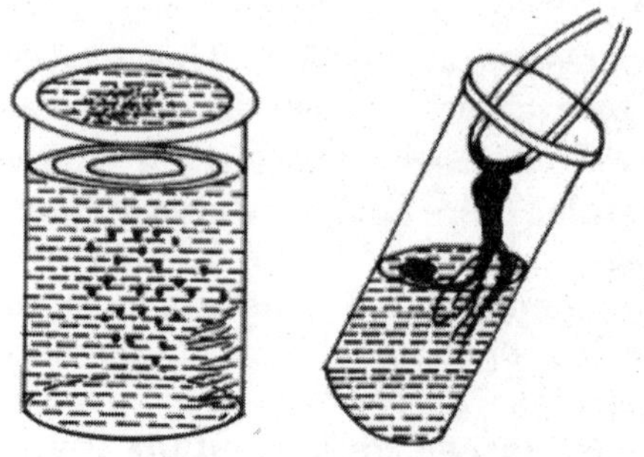

Fig. 7.5. DNA that separates out can be removed by spooling (spool = reel for winding yarn).

2. Cutting of DNA at Specific Locations

The purified DNA molecules are cut with restriction enzyme by incubating at optimal conditions which are specific to that enzyme. Agarose gel electrophoresis is used to check the progression (reunion) of a restriction enzyme digestion. Since DNA is a negatively charged molecule, it moves towards anode (the positive electrode). The vector DNA is also cut with the same restriction enzyme so as to get complementary sticky ends.

3. Separation and Isolation of DNA Fragments

When DNA is treated by restriction endonucleases, fragments of DNA are formed. These fragments are separated by a technique called **gel electrophoresis.** Electrophoresis is a technique of separation of charged molecules under the influence of an electric field so that they migrate in the direction of electrode bearing the opposite charge through a base material called matrix. This technique was developed by **A. Tiselius** in 1937. The base material or matrix through which molecules travel may be polyacrylamide (Polyacryline Gel Electrophoresis or PAGE) or agarose (Agarose Gel Electropnoresis or AGE). Nowadays agarose is the most commonly used matrix which is a natural polymer extracted from sea weeds. Agarose dissolves in hot water, which on cooling forms double helices of thick filaments. The latter become cross linked to form the gel. Since DNA fragments are negatively charged molecules, they can be separated by forcing them to move towards the anode under an electric field through the matrix. The DNA fragments separate (resolve) according to their size through the pores of agarose gel. Hence the smaller the fragment size, the farther it moves.

The DNA fragments can be visualised after staining with **ethidium bromide** and exposing to UV radiation. Pure DNA cannot be seen in the visible light and without staining. DNA fragments appear as bright orange coloured bands in the ethidium bromide stained gel exposed to UV light (Fig. 7.6). The separated bands of DNA are cut out from the agarose gel and extracted out from the gel piece. The process of separation and extraction of DNA fragments from two different sources are used in the formation of recombinant DNA by joining them with the help of cloning vectors.

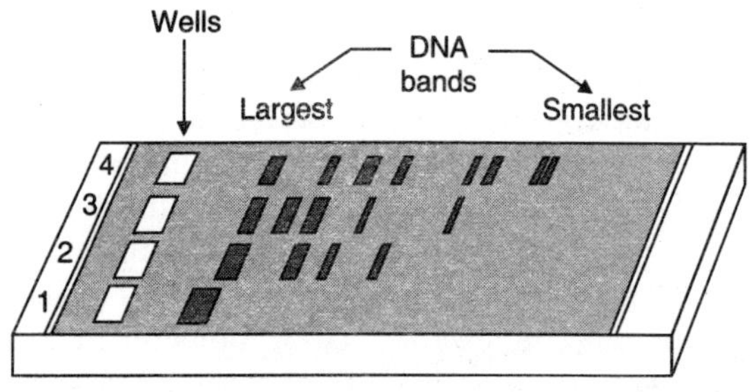

Fig. 7.6. A typical agarose gel electrophoresis showing migration of undigested (lane 1) and digested set of DNA fragments (lane 2 to 4).

4. Ligation of DNA Fragment into a Vector

The joining of DNA involves several processes. After having cut the source DNA as well as the vector DNA with a specific restriction enzyme, the cut out 'gene of interest' from the source DNA and the cut vector with space are mixed and DNA ligase is added. This results in the preparation of recombinant DNA.

5. Transfer of Recombinant DNA into the Host (Host for Transformation with Recombinant DNA)

The cell which is capable of taking up alien DNA is called **competent host.** A competent host is essential for transformation with recombinant DNA. Since DNA is a hydrophillic molecule, it cannot pass through membranes so the bacterial cells must be made competent to take up DNA. It is achieved by treating the bacteria with a specific concentration of a divalent cation, such as calcium, which increases the efficiency with which DNA enters the bacterium through pores in its cell wall. Recombinant DNA can then be forced into such cells by incubating the cells with recombinant DNA on ice, followed by placing them briefly at 42°C (heat shock), and then putting them back on ice. This makes the bacteria to take up the recombinant DNA.

Other Methods of Introducing DNA into Host Cells. Following are some other methods to introduce alien DNA into host cells.

(*i*) **Micro-injection.** In this method recombinant DNA is directly injected into the nucleus of animal cell using micro-needles or micro-pipettes. Using this method, **Jeffey S. Chamberlain** *et.al.*, (1993), of Human Genome Centre, Michigam University, U.S.A., cured mice suffering with a neuromuscular disease which is like muscular dystrophy of humans.

(*ii*) **Biolistics or Gene Gun.** This method was first developed by **Standford** *et.al.*, at Cornell University, U.S.A. in 1987. In this method, microparticles (1–2 μm in size) of gold or tungsten coated with DNA are bombarded with high velocity into the plant cells, using a helium pressure particle gun device (Fig. 7.7). Although this technique was developed for plant, yet it is also used to insert genes into animal cells that promote tissue repair near wounds.

(*iii*) **Disarmed Pathogen Vectors.** This method uses 'disarmed pathogen' (transformed pathogen incapable of causing disease) vectors. These pathogen with recombinant DNA vector, when allowed to infect the cell, transfer the recombinant DNA into the host.

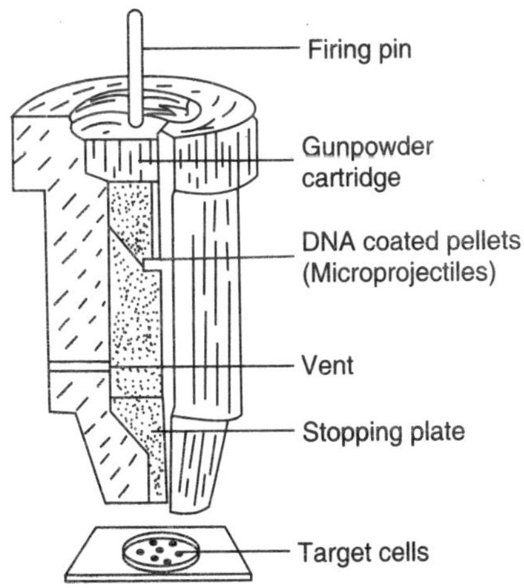

Fig. 7.7. Gene gun.

6. Amplification of Gene of Interest Using PCR

The process of making many copies of a gene is called amplification of gene. It is achieved through a technique called **Polymerase Chain Reaction (PCR).** PCR was originally invented by **Kary Mullis** (1985). It is just like a photocopying a document. In polymerase chain reaction, multiple copies of the gene (DNA) of interest is synthesised *in vitro,* using two sets of primers (small chemically synthesised oligonucleotides that are complementary to the regions of DNA) and the enzyme DNA polymerase. This technique generates upto billion copies of the desired DNA (or RNA) segment within few hours. The PCR amplification cycle involves following three steps.

(*i*) ***Denaturation.*** The target DNA is heated to a high temperature (usually 94°C). It results in separation of the two strands. Each single strand of the target DNA acts as a template for DNA synthesis.

(*ii*) ***Annealing.*** In this step, two small chemically synthesised oligonucleotides (that are complementary to the regions of DNA) primers anneal (or hybridise) to each of the single stranded template DNA.

(*iii*) ***Primer Extension (Polymerization).*** In this step a thermostable enzyme *Taq DNA Polymerase* (isolated from a bacterium, *Thermus aquaticus*) is used to extend primers using the nucleotides and the genomic DNA as template. Synthesis of new DNA strand begins in between the primers, using deoxyribonucleotide triphosphates and Mg^{2+}. The optimum temperature for this polymerization is kept at 72°C. The *Taq DNA Polymerase* remain active during the high temperature induced denaturation of double stranded DNA. If the process of replication of DNA is repeated many times, the segment of DNA can be amplified to approximately billion times (*i.e.,* 1 billion copies are made).

The next PCR amplification cycle begins as soon as all the stages of previous cycle end. During PCR operation, the extension product of one cycle serve as a template for subsequent cycles and each time the amount of DNA doubles. Thus, a single template molecule of DNA generates 2^n molecules at the end of n cycles (Fig. 7.8).

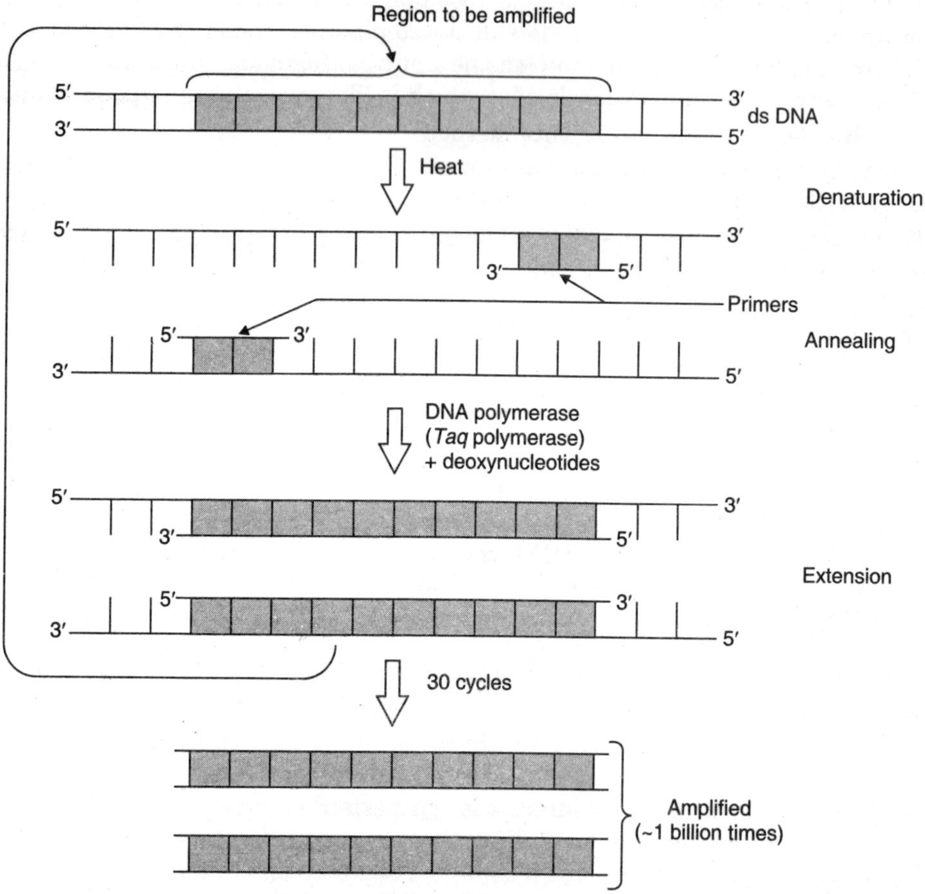

Fig. 7.8. Polymerase chain reaction (PCR): Each cycle has three steps: (*i*) Denaturation; (*ii*) Primer annealing; and (*iii*) Extension of primers.

The amplified fragment if desired can now be used to ligate with a vector for further cloning.

7. Insertion of Recombinant DNA into the Host Cell/Organism

The ligated DNA is introduced into the recipient cells by making them 'competent' to receive (take up) DNA present in its surrounding. Thus, if a recombinant DNA bearing gene for resistance to an antibiotic (*e.g.*, amplicillin) is transferred into *E. coli* cells, the host cells become transformed into ampicillin resistant cells. If the transformed cells are spread on agar plates containing ampicillin, only transformants will grow, untransformed recipient cells will die. Therefore, due to ampicillin resistance gene, one is able to select a transformed cell in the presence of ampicillin. In this case, the ampicillin resistance gene is called as **'selectable marker'.**

8. Obtaining the Foreign Gene Product

Most of the recombinant technologies are aimed to produce desirable proteins. When a recombinant DNA is transferred into a bacterial, plant or animal cell, the alien DNA is multiplied. The foreign gene gets expressed under appropriate conditions. It is important to know the technological details and optimum conditions under which a foreign gene is expressed in host cells.

After cloning the gene of interest, one has to maintain the optimum conditions to induce the expression of the gene and production of target protein on a large scale. If any protein encoding gene is expressed in a heterologous host, it is termed as a **recombinant protein.** The cells containing cloned genes of interest may be grown on a small scale in the laboratory. The cultures may be used for extracting and purifying the desired protein. The cells can also be multiplied in a continuous system (= continuous process), where the used medium is taken out from one side and fresh medium is added from the other side to maintain the cells in their physiologically most active log or exponential phase: *i.e.,* phase of rapid multiplication of cells. Through this type of culturing method, large quantity of biomass is produced which gives rise to higher yields of desired protein.

Since, small volume cultures cannot yield the products in large quantities. Therefore, large steel vessels called **bioreactors** are employed to obtain the products in large quantities. In a bioreactor large volumes (100–1000 litres) of culture can be processed. Hence, bioreactors are like vessels in which raw materials are biologically converted into specific products, individual enzymes, etc., using microbial, plant, animal or human cells. A bioreactor provides the optimal conditions for achieving the desired product by providing optimum growth conditions such as substrate, salts, vitamins, temperature, pH, oxygen etc. The bioreactors are of different types. However, the stirring type is the most commonly used bioreactor.

A stirring type bioreactor usually cylindrical or has a curved base so as to facilitate the mixing of the reactor contents. It has a stirrer which facilitates even mixing and oxygen availability throughout the bioreactor. The bioreactor has a number of systems such as an agitator system, an oxygen delivery system, a foam control system, a temperature control system, pH control system etc. It also has sampling ports, so that small volumes of culture can be taken out periodically (Fig. 7.9).

9. Downstream Processing

The process of separation and purification of a biosynthetic product is called downstream processing. After the formation of a biosynthetic product, it is subjected to a series of processes before it is ready for marketing as a finished product. These processes include separation and

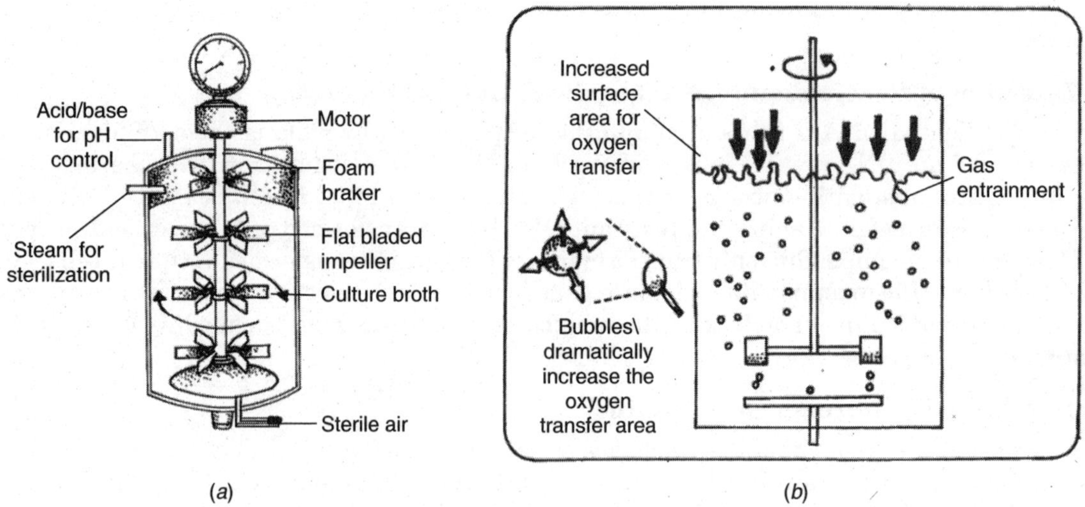

Fig. 7.9. (*a*) Simple stirred-tank bioreactor and (*b*) Sparged stirred-tank bioreactor through which sterile air bubbles are sparged.

purification of the product. The product has to be formulated with suitable preservatives. In case of a drug, such formulation has to undergo through clinical trials. A proper quality control testing is also required for each product. Different type of downstream processing and quality control testing are employed for different type of products.

Genetically Modified Foods

The techniques of modern biotechnology are becoming increasingly important part of the overall effort to improve methods of food production, and to increase the variety and quality of foods. Modern biotechnology has potential applications in the production of food, food processing and also in the assurance of food quality and safety. The foods produced through the techniques of biotechnology such as genetic engineering are called **genetically modified foods.**

Although many genetically modified food crops are under development, very few have so far reached the supermarket shelf. The first genetically modified whole food to reach the shops, was the Flavr Savr^{TM} tomato in USA, and the first product in the UK was tomato puree. Nearly 60 per cent of processed foods contain soy or soy derivatives (such as starch proteins or oils) are being produced from genetically modified soy beans.

Modern biotechnology is also being used to improve micro-organisms, processing aids (*e.g.* enzymes) or ingredients for use in the production and processing of food. The enzyme *chymosin*, used in the production of 'vegetarian cheese' is an example of this type of genetically modified product.

REVIEW EXERCISES

MULTIPLE CHOICE TYPE QUESTIONS

1. The enzyme used in polymerase chain reaction is
   (*a*) Taq polymerase (*b*) RNA polymerase
   (*c*) Ribonuclease (*d*) Endonuclease.

2. PCR (polymerase chain reaction) was discovered by
   (*a*) Wilmut (*b*) A. Jeffreys
   (*c*) Eithoven (*d*) Karry Mullis.
3. PCR method is useful for
   (*a*) Monoclonal antibody production
   (*b*) Hybridoma production
   (*c*) Amplification of DNA for forming billions of copies of itself
   (*d*) All of these.
4. Restriction enzymes are
   (*a*) Not always required in genetic engineering
   (*b*) Essential tool in genetic engineering
   (*c*) Nucleases that cleave DNA at sepcific sites
   (*d*) Both (*b*) and (*c*)
5. Identify the plasmid from the following
   (*a*) *Eco* RI (*b*) pBR 322
   (*c*) *Alu* I (*d*) *Hind* II.
6. Genetic engineering is
   (*a*) making artificial genes
   (*b*) hybridisation of DNA of one organism to that of others
   (*c*) production of alcohol by using micro organisms
   (*d*) making artificial limbs, diagnostic instruments such as ECG, EFG etc.
7. Plasmids
   (*a*) are circuler protein molecules (*b*) are required by bacteria
   (*c*) are tiny bacteria (*d*) confer resistance to antibiotics.
8. The chemical knives of DNA are
   (*a*) Ligases (*b*) polymerases
   (*c*) Endonucleases (*d*) Transcriptases.
9. When genotype of an organism is improved by the addition of foreign genes the process is called
   (*a*) Biotechnology (*b*) Tissue culture
   (*c*) Genetic engineering (*d*) Genetic diversity.
10. There is a restriction endonucleas called *Eco* RI. What does *co* part in it stand for ?
   (*a*) *coli* (*b*) colon
   (*c*) coclom (*d*) coenzyme.

OBJECTIVE TYPE QUESTIONS

1. Fill in the blanks with suitable words :
   (*i*) Specific DNA sequence in a chromosome which is responsible for initiating replication is called ....
   (*ii*) *Hind* II always cut DNA molecule at a particular point by recognising a specific sequence of .... bas pairs.
   (*iii*) To isolate DNA from a fungal cells, the later are treated with ....
   (*iv*) The group of letters that form the same words when read both forward and backward are called ....
   (*v*) .... is a technique by which recombinant DNA is directly injected into the nucleus of the animal cell.

2. Match the items of Column I with the items of Column II.

Column I	Column II
(*i*) Lysozyme	(*a*) Agarose
(*ii*) Tiplasmid	(*b*) *Thermus aquaticus*
(*iii*) Gel electrophoresis	(*c*) α-galactosidase
(*iv*) Taq DNA Polymerase	(*d*) Bacterial cells
(*v*) Insertional inactivation	(*e*) *Agrobacterium tumifaciens*

3. Give the scientific terms for the following :
   (*i*) Nucleotide sequence in DNA responsible for initiating replication.
   (*ii*) The structure used to transfer the piece of DNA attached to it into a host cell.
   (*iii*) Separation of DNA fragments under an electric field through a matrix.
   (*iv*) Extraction of DNA fragments from agarose gel after electrophoresis.
   (*v*) The technique by which multiple copies of a gene/DNA are synthesised.

TEST QUESTIONS

1. Who is known as 'Father of Genetic Engineering' ?
2. What are molecular scissors ?
3. Name the enzyme which cut DNA molecules into fragments with sticky ends.
4. Differentiate between exonucleases and endonucleases.
5. Explain, how bacteria are made capable to take up recombinant DNA ?
6. Explain, how is recombinant DNA constructed ?
7. What are disarmed pathogen vectors ? In what way they are useful in recombinant DNA technology.
8. What is PCR ? Describe its various steps.
9. What is gel electrophoresis ? Write its application in recombinant DNA technology.
10. Describe the tools of genetic engineering.
11. Describe the essential features of a cloning vector.
12. Explain, how are selectable markers used in selection of a transformant ?

8 Gene Sequencing and Human Genome Project

The identity of an organism or an individual is due to its genetic make up. The genetic make up of the organism/individual lies in its DNA sequences. The differences between two individuals are due to differences in the nucleotide sequences of their DNAs, at least at some places. These assumptions led to the quest of finding out the complete DNA sequence of human genome. Advances in genetic engineering techniques made it possible to isolate and clone any piece of DNA and determine nucleotide sequences of these fragments. In 1990, U.S. Department of Energy and National Institute of Health, started the project of sequencing human genome called **Human Genome Project (HGP)**. During the early years of the HGP, the Welcome Trust (UK) became a major partner. Later on Japan, France, Germany, China and others joined the project. By 1993, HGP became an established international effort.

Human Genome Project was called a **mega project** because of its higher magnitude and requirements. HGP is involving a lot of money, experts and most sophisticated and advanced infrastructure. The magnitude of the project can be assessed from its approximate cost. If the cost of sequencing of a bp is 3 dollars, the cost of sequencing of 3×10^9 bp would be 9 billion US dollars. Further, if the obtained sequences were to be stored in typed form in books, and each page of the book contained 1000 letters and each book contained 1000 pages, then 3300 such books would be required to store the information of DNA sequence from a single human cell. This require, the use of high speed computational devices for data storage and retrieval and analysis.

The scientific discipline of mapping, sequencing and analysing genomes is called **genomics.** The field of genomics relies upon **bioinformatics,** which is the management and analysis of biological information stored in data bases. Bioinformatics has immerged as a scientific discipline that encompasses the application of computing science and technology to analyse and manage biological data.

GOAL OF HGP

The goals set for HGP were as follows :

1. Identify all the genes (approximately 20,000–25,000 genes) present in human DNA.
2. Determine the sequence and number of all the base pairs (3 billion bp) present in human DNA.
3. Determine the functions of all the genes.
4. Identify genes causing any disorder.
5. Store this information in data base.
6. Improve tool for data analysis.
7. Transfer related technologies to other sectors, such as industries.
8. Find out the Ethical, Legal and Social Issues (ELSI) that may arise from this project.

HGP officinally started on October 1, 1990, as a 13 year programme. The project was to be completed in 2003. On February 12, 2001, a formal announcement about the completion of the project was made. However, announcement of sequencing of individual chromosomes came in May 2006 with the completion of sequencing of chromosome I. (This was the last of the 24 human chromosome-22 autosomes and X and Y to be sequenced).

Methodologies

There are two major approaches for sequencing the genome : (*i*) Identifying all the genes that expressed as RNA, referred to as **'Expressed Sequence Tags' (ESTs)**, (*ii*) sequencing the whole set of genome that contained all the coding and non-coding sequence, and later assigning different regions in the sequence with functions called **sequence annotation.**

In general, second methodology was followed for HGP. The total DNA from a cell was isolated and broken into random fragments of relatively smaller sizes. These fragments were cloned in suitable host using specialised vectors. The cloning resulted into amplification of each piece of DNA fragment so that it subsequently could be sequenced with ease. The fragments were cloned in suitable hosts like bacteria and yeast, and the vectors used for the insertion of the fragments into the vectors were **BAC** (Bacterial Artificial Chromosomes), and **YAC** (Yeast Artificial Chromosomes). The fragments were sequenced using automated DNA sequencers that worked on the principle of a method developed by **Frederick Sanger** (double Noble laureate also credited for developing method for determination of amino acid sequences in proteins). The sequences were then arranged on the basis of some overlapping regions present in them. This required generation of overlapping fragments for sequencing computer based programmes were used to allign the sequences (Fig. 8.1). The sequences were then annotated and assigned to different chromosomes. The genetic and physical maps on the genome, this was generated using information on polymorphism of restriction endonuclease recognition sites, and some DNA sequences known as micro-satellites.

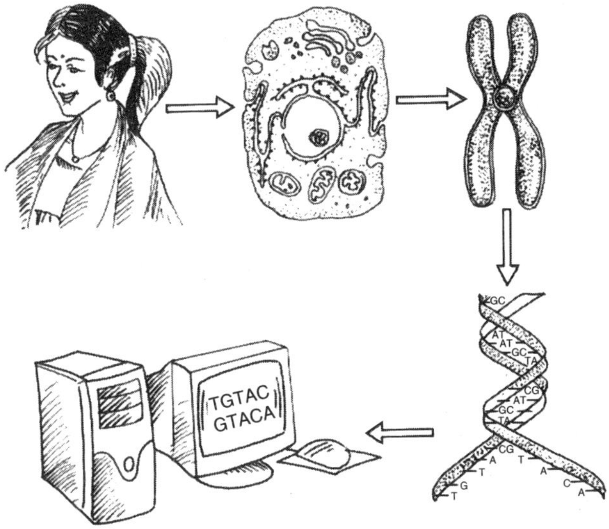

Fig. 8.1. Representative diagram of human genome project.

Salient Features of Human Genome

Some of the important observations of human genome project are as follows.

1. Human genome has 3.1647 billion nucleotide base pairs.

2. The sizes of genes vary greatly; the average gene consists of 3000 base pairs and the largest known human gene (dystrophin or Duchenne Muscular Dystrophy on X chromosome) has 2.4 million (2400 kilo) base pairs. B-globin and insulin genes consist of less than 10 kilobases.

3. The human genome consists of about 30,000 genes much lower than previous estimates of 80,000 to 1,40,000 genes. Human gene count is around the same as that of mouse. Nine-tenth of our genes are identical to that of the mouse. We have more than twice as many genes as fruitfly (*Drosophila melanogaster*) and six times more genes than in bacterium *Escherichia coli* (Fig. 8.2).

4. Chromosome I has the maximum genes (*i.e.,* 2968 genes) and the Y chromosome has the minimum number of genes (*i.e.,* 231 genes).

5. The functions of over 50 per cent of discovered genes are unknown.

6. Less than 2 per cent of the genome codes for proteins.

7. Almost all (99.9 per cent) nucleotide bases are exactly the same in all human beings.

8. Only 0.1 per cent of the human genome (*i.e.,* some 3.2 million nucleotides) represents the variability.

9. Repeated sequences make up a large portion of human genome.

10. Repeated sequences are nucleotide sequences that are repeated many times, sometimes hundred to thousand times. They do not have direct coding functions, but they shed light on chromosome structure, dynamics and evolution.

11. There are about 1.4 million locations where single base DNA differences (called **SNPs—Single Nucleotide Polymorphism** pronounced as **'Snips'**) occur in human genome. Such observations help in finding chromosomal locations for disease associated sequences and tracing human history.

12. Approximately 1 million copies of short 5 to 8 base pair repeated sequence clustered around the centromeres and near the ends of the chromosomes. These sequences are called **'junk DNA'.**

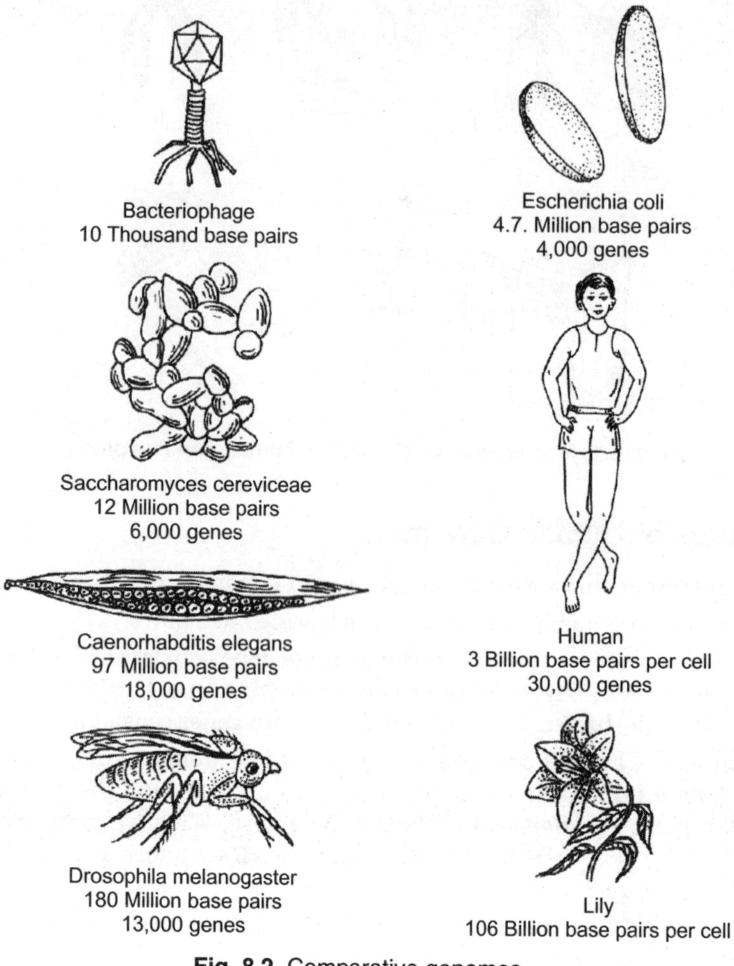

Fig. 8.2. Comparative genomes.

Applications and Future Challenges

The probable applications and future challenges of human genome project are as follows:

1. **Understanding of Biological Systems.** Deriving meaningful knowledge from the DNA sequences will define research through the coming decades leading to our understanding of biological systems. This enormous task will require the expertise and creativity of tens of thousands of scientists from varied disciplines in both the public and private sectors worldwide.

2. **New Approach to Biological Research.** The greatest impacts of having the HGP sequence may be enabling a radically new approach to biological research. The earlier approach was to study one or a few genes at a time. With whole genome sequences and new high throughput technologies, we can approach questions systematically and on a much broader scale. Now researchers can study all the genes in a genome. For example, all the transcripts in a particular tissue or organ or tumor or how tens of thousands of genes and proteins work together in interconnected networks to orchestrate the chemistry of life.

3. **Understanding of Disorders.** Through HGP we will soon be able to have snapshots of more than 1200 genes that are supposed to be responsible for common cardiovascular ailments, endocrine diseases like diabetes, neurological disorder like Alzheimer's disease, deadly cancers and many more. Even defects in a single gene may account for several inherited diseases.

4. **Control of Cancer.** This project aims at to trace genes that will revert cancerous cells to normal.

5. **Health Care.** The human genome sequencing holds promise for a healthier living, designer drugs, genetically modified diets and our genetic identity.

DNA FINGERPRINTING

DNA fingerprinting more technically called **DNA typing** or **DNA profiling** is a method of ascertaining relationship and the identity of a person by means of a DNA fingerprint, which is unique to each individual. It cannot be changed by any known treatment. The ideal way to distinguish an individual from other people would be his or her entire genomic DNA sequence. Thus, DNA fingerprinting can be defined as *'a technique to identify a person on the basis of his / her DNA specificity.'* The technique of DNA fingerprinting was invented by **Sir Alec Jeffreys** (1985, 86) at Leicester University UK. In India DNA fingerprinting technology was started by **Dr. V.K. Kashyap** and **Dr. Lalji Singh.**

Principle of DNA Fingerprinting

As stated earlier that 99.9 per cent of base sequence among humans is the same. Thus, the difference of about 0.1 per cent or 3×10^6 base pair (out of 3×10^9 bp) provide individuality to each human being. If one has to find genetic differences between two individuals or among individuals of a population, sequencing the DNA every time would be a tedious and expensive task. DNA fingerprinting is a very quick way to compare the DNA sequences of any two individuals.

DNA fingerprinting involves indentifying differences in some specific regions in DNA sequence called as **repetitive DNA.** In these sequences, a small stretch of DNA is repeated many times. They can be separated as **satellite DNAs** from the bulk genomic DNA during density gradient centrifugation. Depending on base composition (A : T rich or G : C rich), length of segment, and number of repetitive units, the satellite DNAs can be classified into many categories such as micro-satellite, mini-satellite etc. These sequences normally do not code for any proteins, but they form a large portion of human genome. The satellite DNAs show high degree of **polymorphism** and form the basis of DNA fingerprinting. If a variant at a locus (allelic sequence variation) is present with a frequency of more than 0.01 in human population, it is called **DNA polymorphism.** Variations occur due to mutations. Therefore, *DNA polymorphism is the occurrence of mutations in a population at high frequency.* The probability of such variations would be higher in non-coding DNA sequence, as mutations in these sequences may not have any immediate effect/impact in an individual's reproductive ability. These mutations keep on accumulating generation after generation, and form one of the basis of variability/polymorphism. There is present a variety of different types of polymorphisms ranging from single nucleotide change to very large scale changes. Such polymorphism play an important role in speciation and evolution.

Technique for DNA Fingerprinting

In DNA fingerprinting technique, satellite DNA is used as probe that shows very high degree of polymorphism. It is called as **Variable Number of Tandem Repeats** (VNTR). These are also called **'mini-satellites'**. Each individual inherits these repeats from his/her parents which are used as genetic markers in a personal identity test. For example, a child might inherit a chromosome with six tandem repeats from the mother and the same tandem repeat four times in the homologous chromosome inherited from the father. The half of VNTR alleles of the child resemble that of the mother and half that of the father (Fig. 8.3).

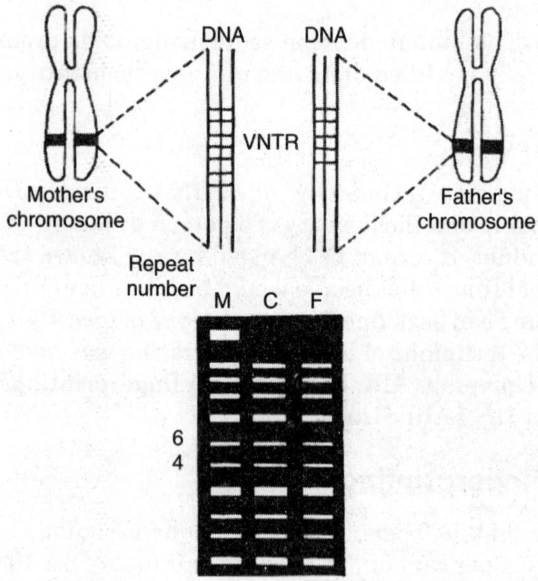

Fig. 8.3. Variable Number Tandem Repeats (M = mother, F = father ; C = child).

The technique, as used earlier, involved southern blot hybridization using radio-labelled VNTR as a probe (Fig. 8.4). DNA fingerprint can be prepared from extremely minute amount of blood, semen, hair bulb or any other cells of the body. It includes :

1. **Isolation of DNA.** DNA is extracted from the nuclei of white blood cells or of spermatozoa or of the hair follicle cells that cling to the roots of hair that have fallen or been pulled out.

2. **Digestion of DNA.** The DNA molecules are first digested with the help of enzyme restriction endonuclease that cut them into fragments. The fragments of DNA also contains the VNTRs.

3. **Separation of DNA Fragments.** The fragments are separated according to size by gel electrophoresis.

4. **Amplification.** If required VNTRs are multiplied by a technique called **Polymerase Chain Reaction** (PCR).

5. **Southern Blotting.** The separated fragments of DNA in the gel are copied/ transferred to synthetic membrane such as nitrocellulose or nylon membrane.

6. **Hybridization Using Labelled VNTR Probe.** Special DNA probes are made in the laboratory. These contain repeated sequences of bases complementary to those on VNTRs. These probes are made radio-labelled with radioactive isotopes. The radio-labelled DNA probes bind to the repeat sequences on the nylon membrane.

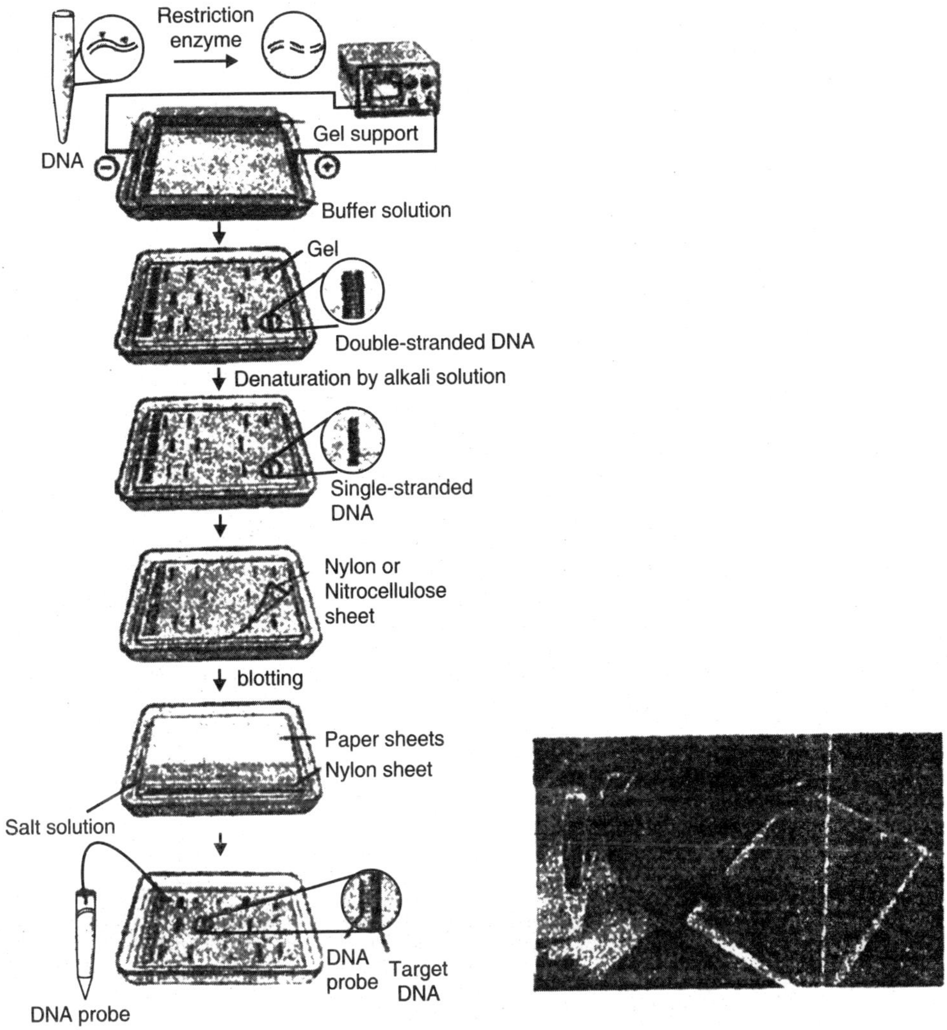

Fig. 8.4. Technique of DNA fingerprinting.

7. **Detection of Hybridized DNA Fragments by Autoradiography.** An X-ray film is exposed to the nylon membrane to mark the places where the radioactive DNA probes have bound to the DNA fragments. These places are marked as dark bands when X-ray film is developed. The dark bands on X-ray film represent the **DNA fingerprints** or **DNA profiles.**

Application of DNA Fingerprinting

1. **Paternity Tests.** The major application of DNA fingerprinting is in determining family relationships. For identifying the true (biological) father, DNA samples of child, mother and possible fathers are taken and their DNA fingerprints are obtained. The prints of child DNA match to the prints of biological parents.

2. **Identification of Criminals.** DNA fingerprinting has now become useful technique in forensic (crime detecting) science, specially when serious crimes such as murders and rapes are involved. For identifying a criminal, the DNA fingerprints of the suspects from blood or hair or semen picked up from the scene of crime are prepared and compared. The DNA

fingerprint of the person matching the one obtained from sample collected from scene of crime can give clue to the actual criminal (Fig. 8.5).

3. **Study of Evolution.** DNA fingerprinting is used for studying evolution. The technology is useful in finding out genetic relationship between different ethenic groups and different races of man.

4. **Identification of Immigrants.** DNA fingerprinting helps to verify, whether a hopeful immigrant is as he or she claims really a close relative of already established resident.

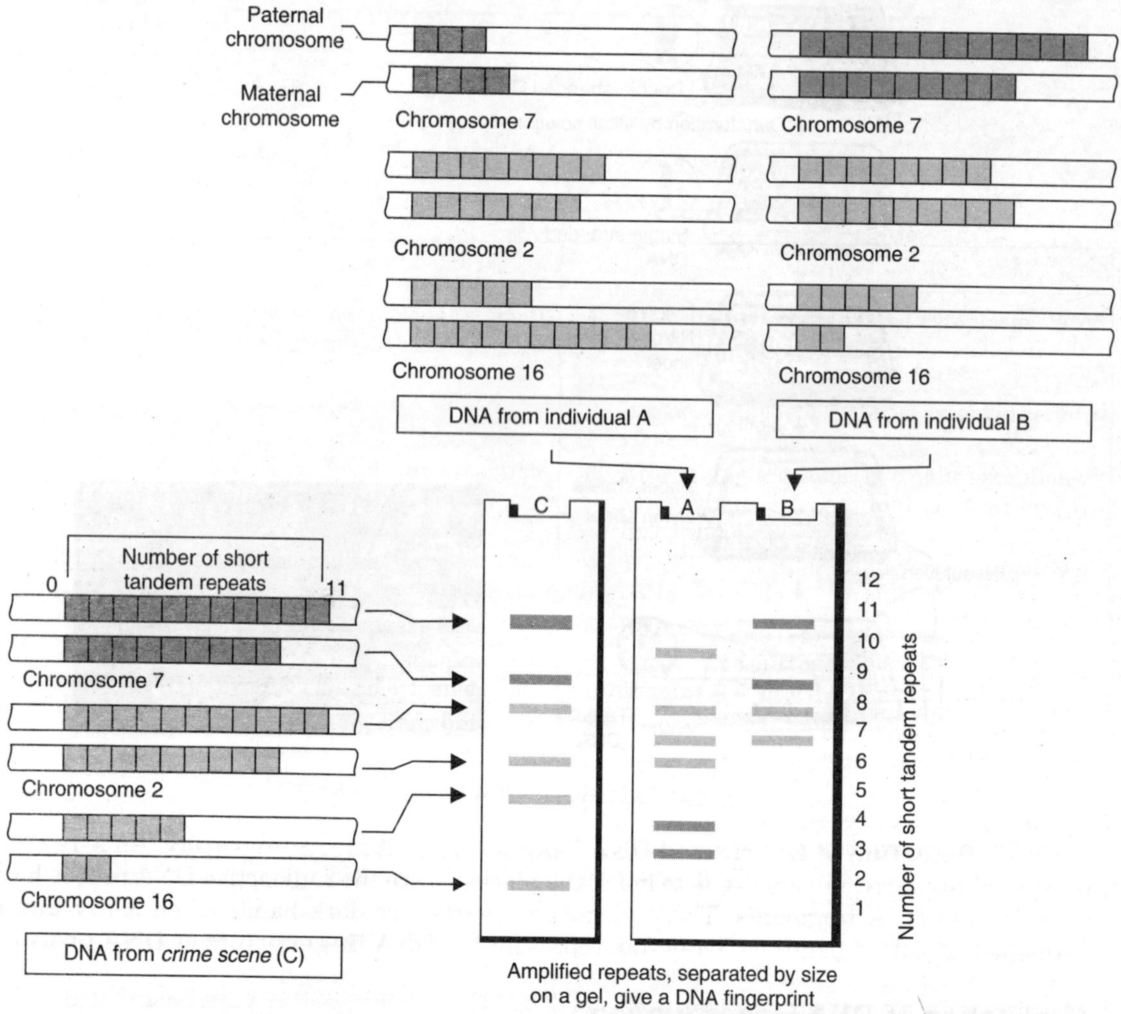

Fig. 8.5. Schematic representation of DNA fingerprinting : Few representative chromosomes have been shown to contain different copy number of VNTR. For the sake of understanding different shade schemes have been used to trace the origin of each band in the gel. The two alleles (paternal and maternal) of a chromosome also contain different copy numbers of VNTR. It is clear that the banding pattern of DNA from crime scene matches with individual B, and not with A.

GENE CLONING

The term 'clone' means different for different purposes. For general purposes, a clone means an exact duplicate copy of an original material. A microbiologist uses the term cloning for production of a population of microbes generated from a single cell. In a biological sense, cloning is the production of an individual organism from a nucleus or a cell. However, the aims and methods differ greatly between the cloning of plants and of higher animals. A biotechnologist particularly a genetic engineer, uses the term cloning for production of multiple and identical copies of the gene (DNA) inside the host cell (usually a bacterium). This kind of cloning is specifically called **gene cloning.** (Fig. 8.6)

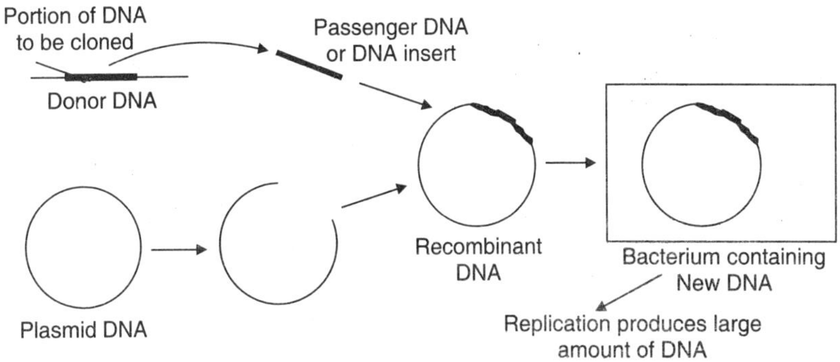

Fig. 8.6. Diagram explaining gene cloning.

Gene Therapy

Gene therapy is the treatment of hereditary disorders by replacing a faulty gene by a normal healthy functional gene. It is a collection of methods that allows correction of a gene defect that has been diagnosed in a child or embryo. In gene therapy, normal genes are inserted into a person's cells and tissues to treat a hereditary defect. Correction of a genetic defect involves delivery of a normal gene into the individual or embryo to take over the function or compensate for the non-functional gene. Gene therapy is being tried for sickle cell anaemia and Severe Combined Immuno-Deficiency (SCID).

SCID is caused due to deletion of the gene for the enzyme *adenosine deaminase* (ADA) in 25 per cent of the cases. These patients have non-functioning T-lymphocytes and so they cannot mount immune responses against invading pathogens. The ideal approach for SCID treatment would be to give patient a functioning ADA that breaks down toxic biological products. The first clinical gene therapy was given in 1990 to a 4 year-old girl with *adenosine deaminase* (ADA) deficiency.

In some children ADA deficiency can be cured by bone marrow transplantation. In others, it can be treated by enzyme replacement therapy, in which functional ADA is given to the patient by injection. But both of these approaches are not completely curative. However, in gene therapy, lymphocytes from the blood of the patient are grown in a culture outside the body. A good copy of human gene encoding this enzyme (ADA cDNA) is introduced into these cells using a retroviral vector. These genetically engineered lymphocytes are subsequently returned to the patient (Fig. 8.7). Since these cells are not immortal, the patient requires periodic infusion of such genetically engineered lymphocytes. However, if the gene isolated from marrow cells producing ADA is introduced into cell at early embryonic stages, the disease could be cured permanently. The functional genes introduced to overcome defective genes are called **therapeutic genes.**

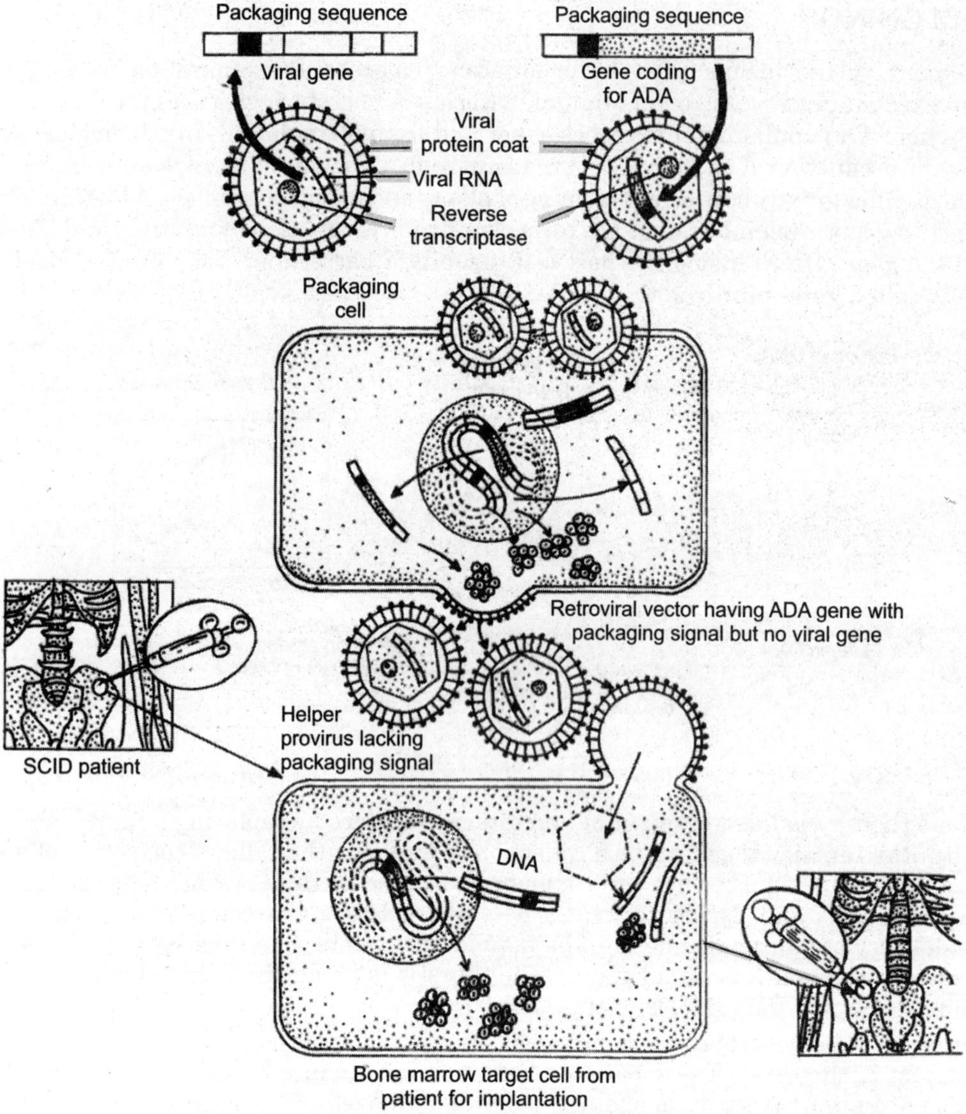

Fig. 8.7. Gene therapy to treat SCID.

Stem Cells

Stem cells are specialised cells that can develop into any type of cells. The cells of early animal and plant embryos have such capability and are called stem cells. A stem cell is a primitive type of cell capable of developing into most of the 220 types of cells found in the human body. The stem cells provide tissue renewal and may be used to grow tissues and organs for replacements.

The stem cells can be cultured from early human embryos. Such stem cells are called **embryonic stem cells.** Stem cells can also be isolated from blood, brain or muscle tissue. There are called **adult stem cells.** The advantage of embryonic stem cell is that they are 'master' cells, and are able to develop into any sort of tissue (such as muscle, nerve tissue, blood cells, heart cells, etc.)

Stem cells display two properties. When divide by mitosis the daughter cells can either.

(*i*) enter a path leading to a fully differentialid cell or.

(*ii*) remain a stem cell, thus ensuring that the pool of stem cells is not used up.

Based on developmental potential, the stem cells are of three types.

1. Totipotent cells. The totipotent cells have the potential to become any type of cell in the adult body and also any type of extra embryonic membrane (*e.g.*, placenta). The fertilised egg and the first four or so cells produced by its cleavage, are totipotent cells. The birth of identical twins, triplets in mammals including humans is due to totipotency of the early embryonic cells. However, in mammals the expression of totipotent stem cells is a misnomner, because these cells fail to meet the second critarion *i.e.*, they cannot make more of themselves.

2. Pluripotent stem cells. These are true stem cells, with a potential to make any differentiated cell in the body, but they cannot contribute to making extra embryonic membrane (derived from trophoblast).

Three types of pluripotent stem cells are found in mammals:

- **Embryonic Stem (ES) cells:** Can be obtained from the inner cell mass (ICM) of the blastocyst (the embryonic stage, when implantation occurs).
- **Embryonic Germ (EG) cells:** Can be isolated from the gonads in aborted foetus.
- **Embryonic Carcinoma (EC) cells:** Can be isolated from taratocarcinomas (a tumor that occasionally develop in a gonad of a foetus).

3. Multipotent stem cells. These are true stem cells but can differentiate into a limited types of cells, *e.g.*, bone marrow contains multipotent cells that give rise to all the cells of the blood but not to other types of cells. Perhaps multipotent stem cells are found in each organ of the body, where they can replace dead or damaged cells.

The most well studied multipotent stem cells are the hematopoietic stem cells present in bone marrow and in the spleen. The formation of blood cells from hematopoietic cell is called **hematopoiesis.** In human beings about one million hematopoietic stem cells produce one billion RBC, one billion platelets, one million T-cells, and the same number of B-cells per kg of body weight.

Personalized Stem Cell Lines

These are stem cells, which an individual get preserved for his/her personal use in future, if required. Such stem cells may be multipotent stem cells from bone marrow. These cells can be cultured to obtain different types of blood cells, and can be used in need without any risk of rejection by the immune system of the body.

REVIEW EXERCISES

MULTIPLE CHOICE TYPE QUESTIONS

1. The scientific descipline of mapping, sequencing and analysing genomes is called

(*a*) Gene mapping (*b*) Genetic engineering

(*c*) Genomics (*d*) biotechnology.

2. The management and analysis of biological information stored in data bases is called

(*a*) DNA sequencing (*b*) Bioinformatics

(*c*) Sequence annotation (*d*) Genomics.

3. The technique of DNA fingerprinting was invested by
   (*a*) Dr. V.K. Kashyap (*b*) Dr. Lalji Singh
   (*c*) Alec Jeffreys (*d*) F. Sanger.
4. In DNA fingerprinting the VNTRs are multiplied through
   (*a*) Gel electrophoresis (*b*) Spooling
   (*c*) PCR (*d*) Elution.
5. The process of transfer of DNA fragments from gel to synthetic membrane is called
   (*a*) Southern blotting (*b*) Western blotting
   (*c*) PCR (*d*) Hybridisation.

OBJECTIVE TYPE QUESTIONS

1. Fill in the blanks with suitable words.
   (*i*) The Human Genome Project (HGP) was initiated by .... in 1990.
   (*ii*) The two major approaches for sequencing the human genome were, Expressed sequence Tags (ESTs) and ....
   (*iii*) The technique to identify a person on the basic of his/her DNA specificity is called ....
   (*iv*) Less than .... per cent of the human genome codes for proteins.
   (*v*) .... sequences make up a large portion of human genome.
2. Match the items in Column I with those in Column II.

Column I	Column II
(*i*) SNPs	(*a*) Identifying all the genes that expressed as RNA
(*ii*) PCR	(*b*) DNA sequences with high degree of polymorphism
(*iii*) VNTRs	(*c*) Single base DNA differences.
(*iv*) ESTs	(*d*) Multiplication of DNA sequences.

TEST QUESTIONS

1. What is human genome project ? Name the organisation that initiated the project.
2. Why was HGP called a megaproject ?
3. Explain the major approaches that was used in sequencing the human genome.
4. Mention the goals of HGP.
5. List the salient features of Human Genome.
6. Explain the steps of DNA fingerprinting that will help in processing of the two blood samples A and B, picked up from the crime scene.
7. Write the full form of VNTR. How is VNTR different from probe ?
8. What are satellite DNA in a genome ? Expalin their role in DNA fingerprinting.
9. What are stem cells ? Give an account of their types and their utility.

UNIT–5

HUMAN REPRODUCTION AND REPRODUCTIVE HEALTH

9 Human Reproduction

Humans reproduce sexually only and are **viviparous** (*i.e.,* give birth to the young ones). They are **unisexual** *i.e.,* male and female sexes are separate. Each sex has reproductive organs (gonads), reproductive ducts and accessory structures. There are male gonads (testes) in males and female gonads (ovaries) in females. They produce gametes and sex hormones. The organs which neither produce gametes nor secrete sex hormones but perform important functions in reproduction are called **secondary sex organs.** The latter include the prostate, seminal vesicles, vasa deferentia and penis in males and the fallopian tubes, uterus, vagina and mammary glands in females. The characters which distinguish the male from the female externally are called **secondary sex characters.**

REPRODUCTIVE EVENTS

In human beings, the major reproductive events include: (*i*) **Gametogenesis.** Formation of gametes *i.e.,* sperms in males and ovum in females (*ii*) **Insemination.** Transfer of sperms into the female genital tract. (*iii*) **Fertilization.** Fusion of male and female gametes, leading to the formation of zygote. (*iv*) **Implantation.** Formation and development of blastocyst and its attachment to the uterine wall. (*v*) **Gestation.** Embryo development or the time from conception to birth. (*vi*) **Parturition.** Delivery of the baby or the process of birth. All these reproductive events occur after **puberty** *i.e.,* when the reproductive organs start functioning. The reproductive events in the males and females differ remarkably. For example, the mammary glands are well developed in the females and rudimentary in males; the females do not develop beard and moustache but males do; the voice in females is pitched higher than in males; in men formation of sperms continues even in old age, but in women the formation of ovum ceases at around the age of 50 years.

THE MALE REPRODUCTIVE SYSTEM

The male reproductive system is located in the pelvis region. It consists of a pair of **testis, genital ducts, accessory glands** and **external genitalia** (Fig. 9.1).

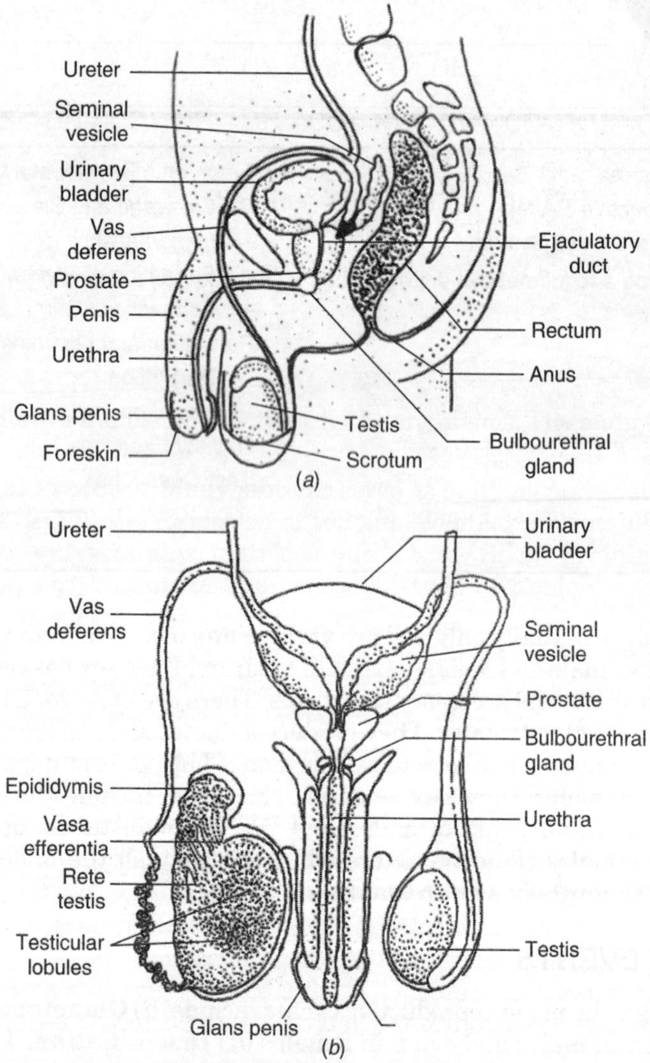

Fig. 9.1. Human male reproductive system; (*a*) side view and (*b*) Front view (part of testis is open to show inner details).

1. Testes

These are a paired oval organs situated outside the abdominal cavity within pouches called **scrotum.** During early foetal life, the testes develop in the abdominal cavity just below the kidney, then they descend into the scrotum. The scrotum remain connected with the abdomen or pelvic cavity by the inguinal canals. Each testis is suspended in the scrotum by **spermatic cord,** which is formed from the spermatic artery, vein and nerve, bound together with connective tissue. Tearing of inguinal tissue by any means may result in the bulging out of a loop of intestine into the scrotum. This condition is known as **inguinal hernia.** A fibrous cord that extends from the caudal end of the testis to the scrotal wall is called **gubernaculum.** The scrotum helps in maintaining the low temperature of the testis (2–2.5°C lower than the normal body temperature), necessary for sperm production. When the body is chilled, the smooth muscle contracts and brings the testes closer to the pelvic cavity. When the temperature drops, movement towards the pelvic cavity allows the testes to absorb heat from rest of the body so that sperm cells do not become chilled.

Each testis is about 4 to 5 cm in length and 2 to 3 cm in breadth. The testis is surrounded by three dense coverings called **tunicae**. These are (*i*) **Tunica vaginalis** is the outer covering of the testis, (*ii*) **Tunica albuginea** is a fibrous covering surrounding the testis situated under the tunica vaginalis, and (*iii*) **Tunica vasculosa** consists of a network of capillaries supported by delicate connective tissue which lines the tunica albuginea. Each testis has about 250 compartments called **testicular lobules.** Each lobule contains one to three highly coiled seminiferous tubules. The wall of each seminiferous tubule is lined by **germinal epithelium.** Majority of cells of this epithelium are cuboidal called **male germ cells (spermatogonia),** but a few large cells called **Sertoli** or **sustentacular cells** (supporting cells). The male germ cells (spermatogonia) grow into **primary spermatocytes,** which undergo meiosis to produce **spermatozoa** (sperms). The developing spermatozoa are nourished by the sertoli cells. The regions outside the seminiferous tubules called **interstitial spaces**. In the interstitial spaces there are present small groups of **rounded interstial** or **Leydig's cells** (named after their discoverer *Franz Von Leydig*). These cells synthesize and secrete testicular hormones called **androgens** *i.e.,* male sex hormones (*e.g.,* **testosterone**). Other immunologically competent cells are also present in the interstitial spaces (Fig. 9.2).

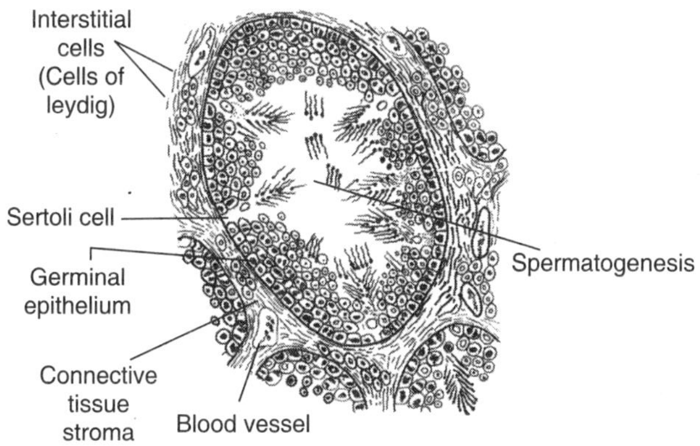

Fig. 9.2. Sectional view of seminiferous tubule.

Testis perform two functions—production of sperms and secretion of male sex hormones.

2. Accessory Ducts

The male sex accessory ducts include **rete testis, vasa efferentia, epididymis** and **vasa deferentia** (Fig. 9.3). The seminiferous tubules are closed at one end but on the other end they join to a network of tubules called **rete testis.** The rete testis open into the vasa efferentia. The vasa efferentia leave the testis and open into epididymis located along the inner side of each testis. At the upper end of the testis, it is called **caput epididymis.** The middle part of the epididymis is termed **corpus epididymis.** The posterior end of the epididymis is known as **cauda epididymis.** The epididymis stores the sperms and also secretes fluid which is considered to nourish the sperms.

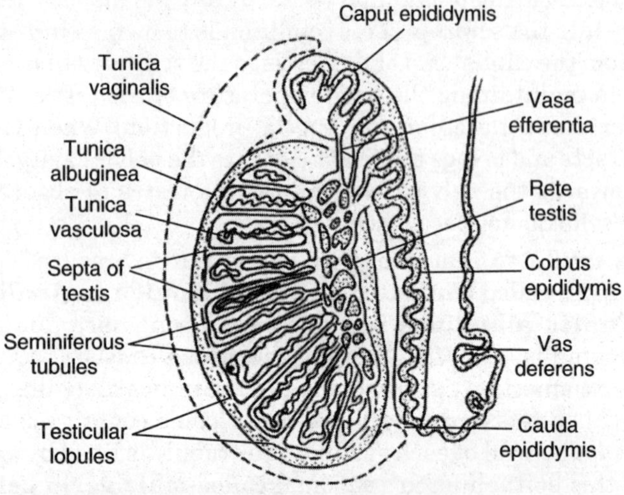

Fig. 9.3. L.S. of mammalian testis and different ducts.

The epididymis leads to vas deferens that ascends to the abdomen and loops over the urinary bladder, where it is joined by the duct from **seminal vesicle** to form the **ejaculatory duct.** The ejaculatory ducts of the two sides pass through the prostate gland and open into the **urethra.** The ejaculatory ducts carry sperms and the secretion of seminal vesicles to the urethra. The urethra originates from the urinary bladder and extends through the penis to its external opening called **urethral meatus.** The urethra provides a common pathway for the flow of urine and semen.

Differences between Vasa Efferentia and Vasa Deferentia

Vasa Efferentia	*Vasa Deferentia*
1. They are thin ducts, which arise from the rete testis.	1. They are comparatively thicker ducts which arise from the cauda epididymis.
2. They vary from 15 to 20 in number.	2. They are only two in number.
3. Their inner lining consists of ciliated cells.	3. Their inner lining has stereocilia.
4. They carry spermatozoa from the rete testis to the epididymis.	4. They carry spermatozoa from cauda epididymis to ejaculatory duct.

3. Penis

The penis is the male external genitalia (male copulatory organ). It consists of a long shaft having an expanded tip called the **glans penis.** It is covered by a loose fold of skin called the **prepuce** or **foreskin**. At the tip of glans penis, there is a slit like opening called the **external urethral orifice** or **meatus** or **urinogenital aperture.** Under the skin, penis contains three columns of erectile tissue—two dorsal **corpora cavernosa** and one ventral **corpus spongiosum**. The corpus spongiosum contains the penile urethra, and also form the glans penis. (Figs. 9.4 and 9.5). During sexual arousal the three bundles of tissue in the penis become engorged with blood. The penis in addition to conducting urine from the body transfers semen into reproductive tract of female during sexual intercourse (coitus).

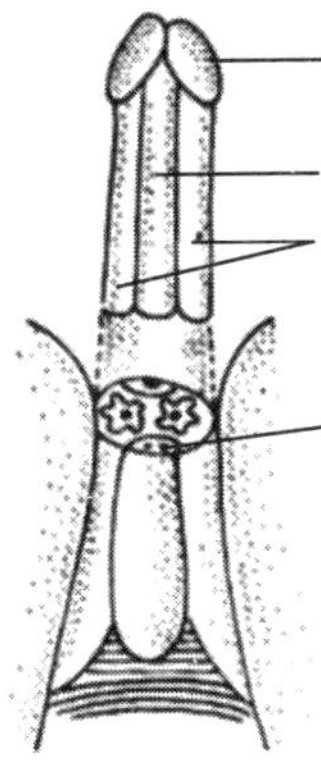

Fig. 9.4. Ventral view of penis.

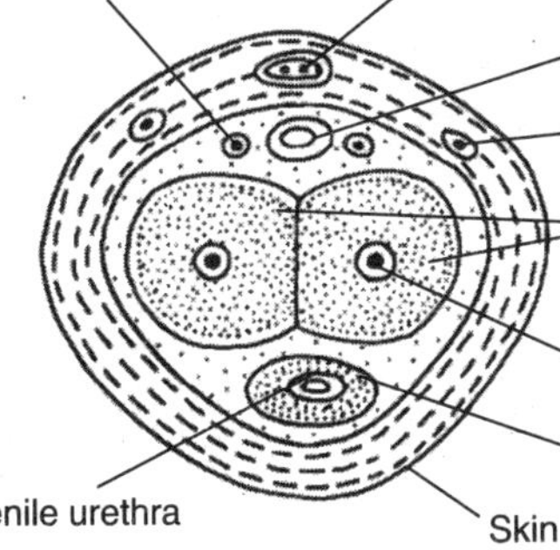

Fig. 9.5. T.S. of penis.

4. Male Accessory Glands

These include a pair of **seminal vesicles**, a **prostate** and a pair of **bulbourethral glands.**

(*i*) ***The seminal vesicles*** are one pair of sac like structures near the base of the bladder. Their ducts join the vasa deferentia to form the ejaculatory ducts. They produce an alkaline secretion which forms 70% of the volume of semen. The secretion of the seminal vesicle contains spermatozoa activating substances such as fructose, citrate, inositol, prostaglandins and several proteins. Fructose is a source of energy for the sperm. Prostaglandins stimulate uterine contractions and thus may help the sperm to be moved towards the female's oviducts, where fertilisation takes place. Some proteins of the secretion help coagulation of semen after ejaculation. Alkalinity of the seminal fluid helps to neutralize the acidic environment of the male urethra as well as that of female reproductive tract which would otherwise inactivate and kill sperms.

(*ii*) ***The prostate gland*** is of a size of a golf ball and surrounds the proximal portion of urethra. It produces a milky, slightly acidic secretion (pH = 6.5) which form 25% of the volume of semen. The secretion of the prostate gland nourishes the spermatozoa and aids in their motality.

(*iii*) ***The bulbourethral glands or cowper's glands*** are present on either side of the proximal portion of urethra. They secrete mucus that helps in the lubrication of penis and linning of urethra.

The secretions of the above glands constitute **seminal plasma** which is rich in fructose, calcium and certain enzymes. Fructose which is produced by the seminal vesicles, is not present anywhere else in the body, provides a forensic test for rape. The presence of fructose in the female's genital tract confirms sexual intercourse.

Semen. Semen is the mixture of spermatozoa and seminal plasma (the secretions of the seminal vesicles, prostate gland and cowper's glands). It is ejected from the penis during ejaculation. A single ejaculate may contain 200 to 300 million spermatozoa (sperms). Semen is slightly alkaline (pH 7.35 to 7.50). Its alkalinity helps to neutralize the acidity of urethra left from the passage of urine and protects the sperms from the acidity of vagina.

THE FEMALE REPRODUCTIVE SYSTEM

The female reproductive system consists of a pair of **ovaries** alongwith a pair of **oviducts, uterus, cervix, vagina, accessory genital glands,** and **external genitalia**

(Fig. 9.6) located in pelvic region, and a pair of **mammary glands** (breast) in thorax region. These parts of the female reproductive system are associated structurally and functionally to support the process of ovulation, fertilisation, pregnancy, birth and child care.

1. Ovaries

Ovaries are the primary female sex organs that produce the female gamete (ovum) and several steroid hormones (ovarian hormones). The ovaries are located one on each side in the upper pelvic cavity. Each ovary shaped like an unshelled almond and is about 2 to 4 cm in length. Each ovary is connected to the pelvic wall and uterus by **ovarian ligament.** The **broad ligament** of the uterus which is itself a part of the parietal peritoneum attaches to the ovary by a double layered fold of peritoneum called the **mesovarium.**

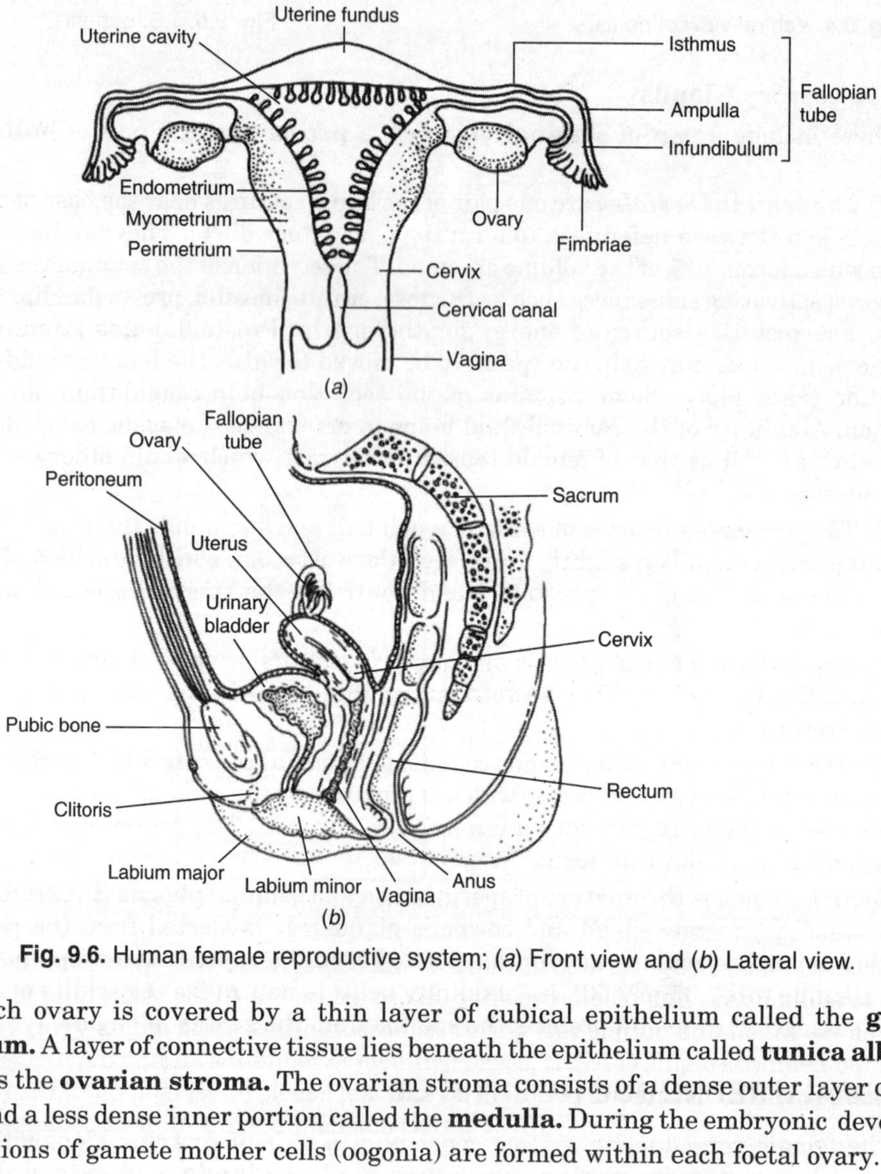

Fig. 9.6. Human female reproductive system; (*a*) Front view and (*b*) Lateral view.

Each ovary is covered by a thin layer of cubical epithelium called the **germinal epithelium.** A layer of connective tissue lies beneath the epithelium called **tunica albuginea.** It encloses the **ovarian stroma.** The ovarian stroma consists of a dense outer layer called the **cortex** and a less dense inner portion called the **medulla.** During the embryonic development stage millions of gamete mother cells (oogonia) are formed within each foetal ovary. No more oogonia are formed and added after birth. The oogonia start division and enter into prophase I of the meiotic division and get temporarily arrested at that stage, and are called **primary oocytes.** Each primary oocyte gets surrounded by a layer of granulosa cells and then called the **primary follicle.** A large number of these follicles degenerate during the phase from birth to puberty. The degenerative process of follicles is called **follicular atresia** and such follicle are known as **atresic follicles.** Therefore, at puberty only 60,000–80,000 primary follicles are left in each ovary. The primary follicles are surrounded by more layers of granulosa cells and called **secondary follicles.** The secondary follicle soon changes into **tertiary follicle** which is characterized by a fluid filled cavity called **antrum.** At this stage, the primary oocyte within the tertiary follicle grows in size and completes its first meiotic division. It is an unequal division resulting in the formation of a large haploid secondary oocyte and a tiny first polar body. The tertiary follicle is then converted into mature follicle called **Graafian follicle** or **ovarian follicle** (Fig. 9.7).

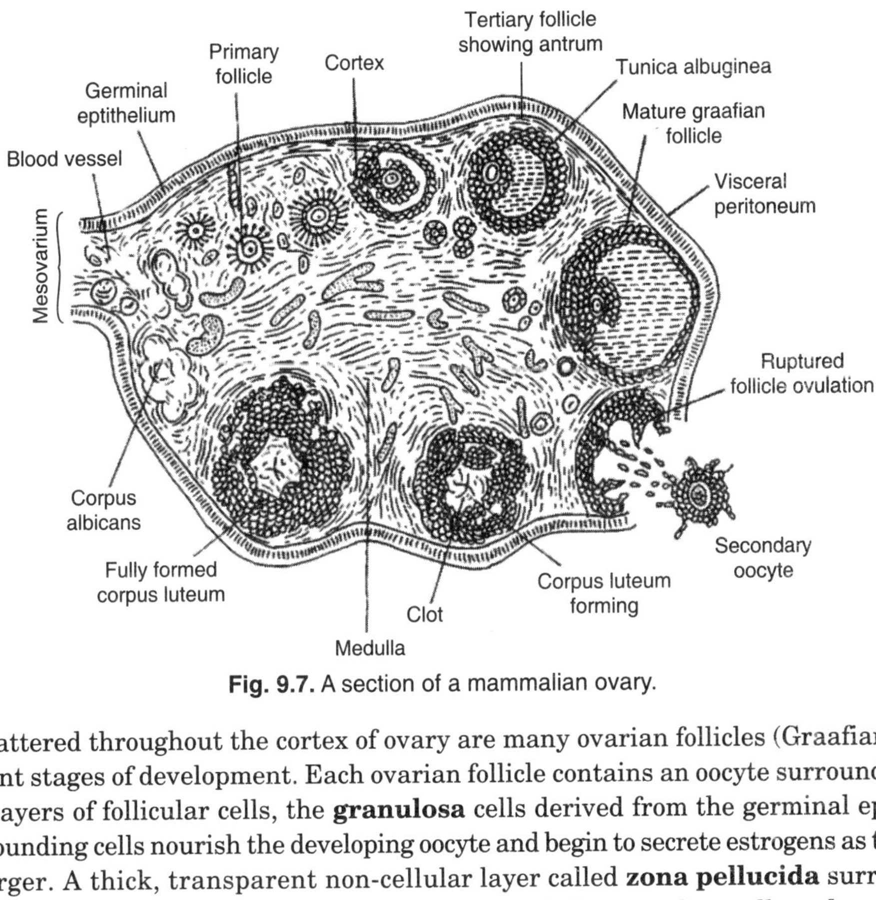

Fig. 9.7. A section of a mammalian ovary.

Scattered throughout the cortex of ovary are many ovarian follicles (Graafian follicles) in different stages of development. Each ovarian follicle contains an oocyte surrounded by one or more layers of follicular cells, the **granulosa** cells derived from the germinal epithelium. The surrounding cells nourish the developing oocyte and begin to secrete estrogens as the follicle grows larger. A thick, transparent non-cellular layer called **zona pellucida** surrounds the oocyte. It is composed of glycoproteins synthesised partly by granulosa cells and partly by the oocyte. Outside zona pellucida is **corona radiata** composed of enlarged follicle cells. The cavity of Graafian follicle, called antrum is filled by a fluid called, the **liquor folliculi.** It is secreted by follicular cells. The mature oocyte is attached to the wall of follicle through a pedicel called **cumulus oophorus.** The membrane granulosa is surrounded by the **theca interna** and **theca externa** (Fig. 9.8).

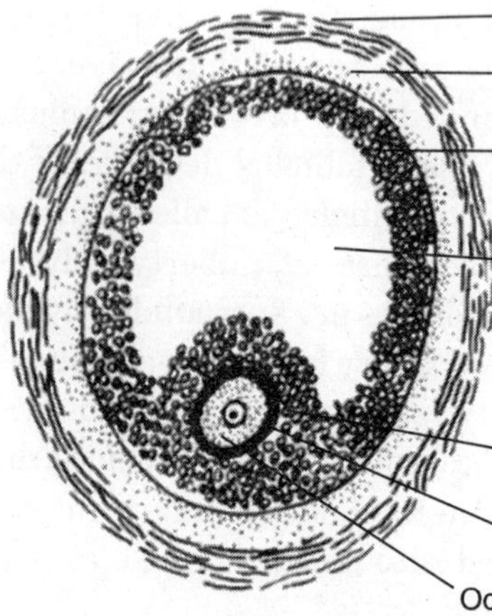

Fig. 9.8. Mature Graafian follicle.

The mature follicle (Graafian follicle) ruptures and releases the ovum (secondary oocyte). This process is called **ovulation.** After ovulation, the granulosa cells and the interstial cells form a mass of large and yellowish conical cells called **corpus luteum.** It serves as a temporary endocrine gland by releasing progesterone and estrogen. Corpus luteum also secretes **relaxin** hormone. If fertilization does not occur, corpus luteum stays for a few days and disappears. The degenerating stage of corpus luteum is called **corpus albicans,** literally meaning white body. Generally, only one ovum is released in each menstrual cycle (average duration, 28 days) by alternate ovaries. Only about 450 ova are produced by the human female over the entire span of her reproductive life which lasts about 40–50 years.

2. The Female Accessory Ducts

The fallopian tubes (oviducts), uterus and vagina constitute the female accessory ducts.

(*i*) ***Fallopian Tubes.*** Each fallopian tube is about 10–12 cm long and extends from the periphery of each ovary to the uterus. The fallopian tube closer to ovary has a funnel like opening called **infundibulum.** The margin of infundibulum possesses finger like projections called **fimbriae** which help in collection of the ovum after ovulation. The infundibulum leads to a wider part of the fallopian tube called **ampulla.** The last part of the fallopian tube that joins the uterus is narrow and is called **isthmus.** The fallopian tubes convey the ovum from the ovary to the uterus. It is achieved by peristalsis. Fertilization of ovum generally takes place at the junction of the isthmus and ampulla (ampullary-isthmus junction) of the fallopian tube.

(*ii*) ***Uterus (womb).*** The uterus is a hollow muscular and inverted pear shaped structure. It lies in the pelvic cavity between the urinary bladder and the rectum, and is attached by ligaments to the pelvic wall. The uterus consists of a body called **corpus** a narrow uterine cavity called **internal Os** and the lower cylindrical structure called **cervix.** The wall of the uterus has three layers of tissue. The outer thin covering is called **perimetrium.** The middle thick layer of smooth muscle is **myometrium.** It shows strong contraction during delivery of the baby. The inner glandular layer that lines the uterine cavity is called **endometrium**. The endometrium undergoes cyclical changes during menstrual cycle.

(*iii*) ***Vagina.*** The cervix projects slightly into an elastic muscular tube called vagina. The vagina is about 10 cm long and opens to the exterior by an aperture called the **vaginal orifice.** The opening of the vagina remains partially covered by a thin ring of a membrane called the **hymen.** It is often torn during the first coitus (intercourse). However, it can also be ruptured due to vigorous physical activities. In some women the hymen persists even after coitus. Therefore, the presence or absence of hymen is not a reliable indicator of virginity. The vagina receives the semen during copulation and serve as the **birth canal** during parturition.

3. External Genitalia (Vulva)

The female external genitalia include mons pubis, labia majora, labia minora, hymen and clitoris (Fig. 9.9). **Mons pubis** is a cushion of fatty tissue covered by skin and pubic hair. The **labia majora** are fleshy folds of tissue which extend down from the mons pubis and surround the vaginal opening. The **labia minora** are smaller folds which lie under the labia majora. Anteriorly, labia minora merge together to form a small eractile organ called **clitoris.** It is homologous to the male's glans penis.

Posteriorly, the labia minora are fused together to form the **fourchette**. They also contain numerous sebaceous glands.

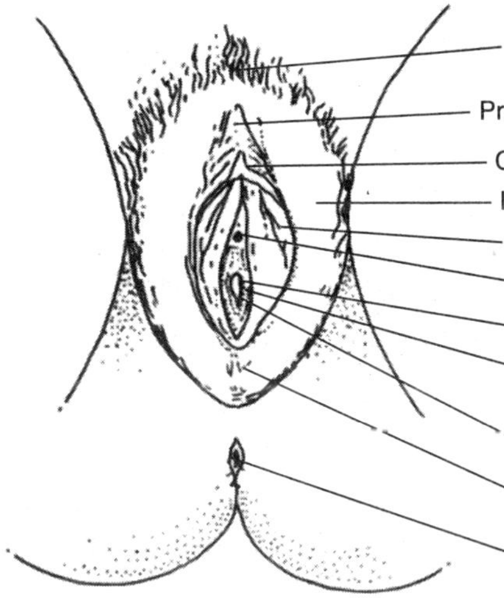

Fig. 9.9. The external genitalia in human female.

4. Accessory Genital Glands

In human female, there are two types of **vestibular glands.** These are *(i)* **Lesser vestibular glands (paraurethral glands** or **glands of skene)** are numerous minute glands that are present on either side of the urethral orifice. These glands are homologous to the male prostate and secrete mucus. *(ii)* **Greater vestibular glands (Bartholin's glands)** are paired glands situated one on each side of the vaginal orifice. These glands are homologous to the bulbourethral (cowper's) glands of the male and secrete viscid fluid that supplements lubrication during sexual intercourse.

5. Mammary Glands (Breasts)

The mammary glands are modified sweat glands, which remain underdeveloped until puberty. At puberty they begin to develop under the influence of estrogen and progesterone hormones. Externally, each breast has a projection, the **nipple** surrounded by a circular pigmented area of skin (deep pink to light brown) called **aerola.** At the surface of aerola there are numerous sebaceous glands called aeolar glands. The mammary glands (breasts) are paired structures that contain glandular tissue and variable amount of fat (Fig. 9.10). The glandular tissue of each breast consists of 15–20 lobes called mammary lobes. Each lobe contains clusters of milk secreting glands called **alveoli.** The milk secreted by the alveoli is stored in the cavities (lumens) of alveoli. The alveoli open into **mammary tubules.** The tubules of each lobe join to form **mammary duct.** Near the nipple, mammary ducts expand to form **mammary ampulla (lactiferous sinuses),** where some milk may be stored before going to the lactiferous duct. The lactiferous ducts carry milk from the lobes to the exterior during the suckling by the baby.

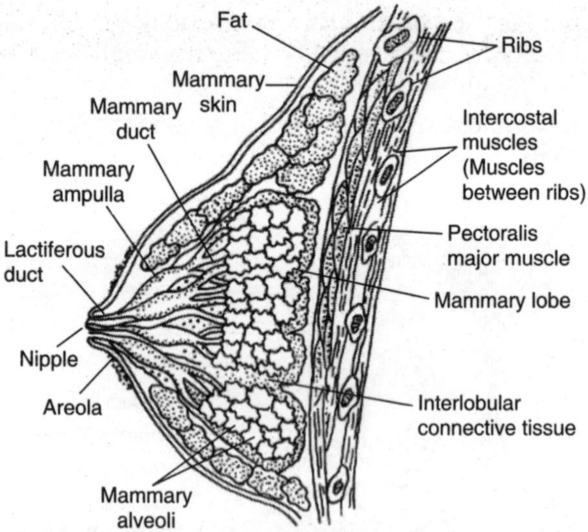

Fig. 9.10. Female's breast in saggital section.

The oviduct, uterus, vagina, clitoris, the accessory genital glands and mammary glands constitute the secondary sexual organs of human female.

Onset of Puberty in the Human Females. In human females, puberty occurs at the age of 11–13 years. At this time, the pituitary gland starts secreting Follicle Stimulating Hormone (FSH). The FSH induces the development of ovaries, which in turn produce the hormone estrogen. This hormone is responsible for the development of external genitalia, breasts, pubic hair and the feminine shape (*i.e.,* widening of the pelvis and deposition of fat in thighs, buttocks and face).

GAMETOGENESIS

The process of formation of gametes in gonads (testes and ovaries) is called gametogenesis. It include **spermatogenesis** (formation of sperms) and **oogenesis** (formation of ova) in the testes and ovaries respectively. The gametes differ from all other cells (somatic cells) in that their nuclei contain only half the number of chromosomes found in the nuclei of somatic cells.

Both spermatogenesis and oogenesis comprise similar phases of sequential changes *viz.* (*i*) multiplication phase, (*ii*) growth phase, and (*iii*) maturation phase.

Spermatogenesis

The process of sperm formation is called spermatogenesis. It takes place in the seminiferous tubules of testis (Fig. 9.11). The seminiferous tubules are lined by **germinal epithelium.** The majority of cells of germinal epithelium are cuboidal and are called **primary** or **Primordial Germ Cells** (PGCs). Some cells of the germinal epithelium are tall and somatic, called **Sertoli cells** (= nurse cells). Spermatogenesis includes formation of spermatids (products of meiosis) and formation of spermatozoa (differentiation of sperms).

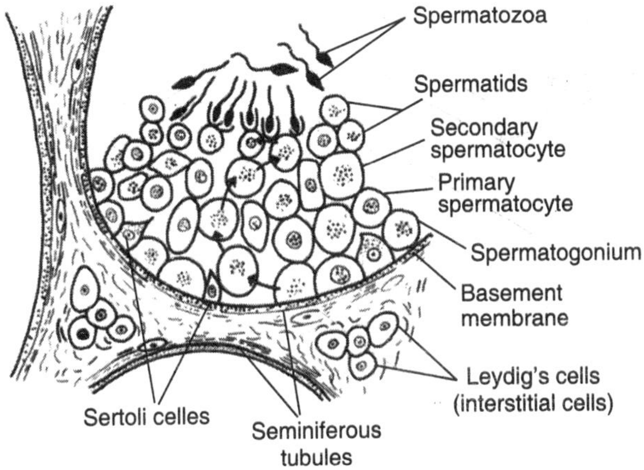

Fig. 9.11. A part of transverse section of mammalian testis.

1. Formation of Spermatids. It completes in following three phases:

(*i*) ***Multiplication Phase.*** At sexual maturity, the undifferentiated primordial germ cells (spermatogonial cells) of seminiferous tubules divide several times by mitosis to produce a large number of **spermatogonia.** (sing. spermatogonium). Each spermatogonium is diploid and contains 46 chromosomes. Some of them continue as stem cells or **type A spermatogonia,** others differentiate into **type B spermatogonia.** Type A spermatogonia serve as stem cells which divide to form type A spermatogonia and type B spermatogonia. Type B spermatogonia act as precursors of sperms.

(*ii*) ***Growth Phase.*** Type B spermatogonia grow in size by obtaining nourishment from the nursing cells and are called **primary spermatocytes.**

(*iii*) ***Maturation Phase.*** Some primary spermatocytes periodically undergo meiosis. The first meiotic division or meiosis I (reductional division) results in the formation of two equal haploid cells (with 23 chromosome each) called **secondary spermatocytes.** The secondary spermatocytes undergo second meiotic division or meiosis II (equational division) to produce four haploid **spermatids** (Fig. 9.12).

2. Formation of Spermatozoa from Spermatids (Spermiogenesis). The transformation of spermatids into spermatozoa is called **spermiogenesis** or **spermateliosis.** It involves, the formation of acrosome, condensation and elongation of nucleus, development of the flagellum and the loss of redundant substances of the cytoplasm and its organelles. Ultimately, each spermatid transforms into a spermatozoan. After spermiogenesis, sperm heads become embedded in the **Sertoli** cells and finally released from the seminiferous tubules by the process called **spermiation.** An adult produces over 10^{12} to 10^{13} spermatozoa each day. These gradually move into the epididymis and the first portion of vasa deferentia, where they undergo maturation and are stored.

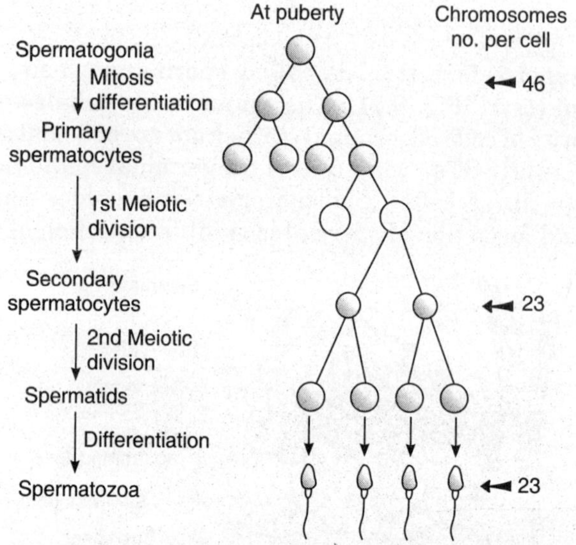

Fig. 9.12. Stages in spermatogenesis (diagrammatic).

Differences between Spermatogenesis and Spermiogenesis

Spermatogenesis	*Spermiogenesis*
1. It is the process of formation of spermatozoa from Primordial Germ Cells (PGCs).	1. It is the process of transformation of spermatids into spermatozoa.
2. It involves the formation of spermatids and their transformation into spermatozoa.	2. It is the part of spermatogenesis.
3. It consists of multiplication phase, growth phase, maturation phase and differentiation.	3. It consists of only differentiation phase.
4. A spermatogonium produces four spermatozoa (sperms).	4. One spermatid produces only one spermatozoa.

Hormonal Control of Spermatogenesis. Spermatogenesis begins at the age of puberty due to significant increase in the secretion of gonadotropin releasing hormone (GnRH) from hypothalamus. The increased levels of GnRH stimulates the anterior lobe of pituitary gland to secrete two gonatotropins—Luteinising Hormone (LH) and Follicle Stimulating Hormone (FSH). LH stimulates leydig cells (interstitial cells) in the testes to synthesise and secrete androgens. Testosterone is the principal androgen. Testosterone is essential for sperm formation. FSH stimulates sertoli cells to secrete two factors—**Androgen Binding Protein** (ABP) that concentrates testosterone in the seminiferous tubules and a protein hormone **inhibin** which suppresses FSH synthesis. FSH acts directly on spermatogonia to stimulate sperm production. The level of testosterone is under negative feedback control. A rising level of testosterone suppresses the release of GnRH from the hypothalamus (Fig. 9.13).

Structure of Spermatozoa. The sperms are microscopic and motile cell. Each sperm is composed of four parts—a head, a neck, a middle piece and a tail (Fig. 9.14). A plasma membrane covers the whole body of sperm.

(*i*) *Head* is the enlarged end of the sperm, containing an elongated haploid nucleus. The anterior of the nucleus is covered by a cap like structure called **acrosome.** The acrosome contains enzymes **sperm lysins** or **hyaluronidases,** which are used to contact and penetrate the ovum at the time of fertilization.

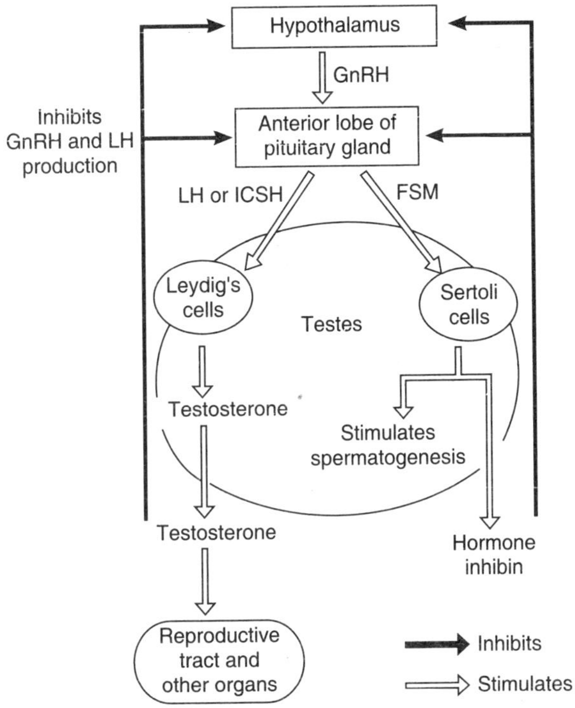

Fig. 9.13. Hormonal control of spermatogenesis.

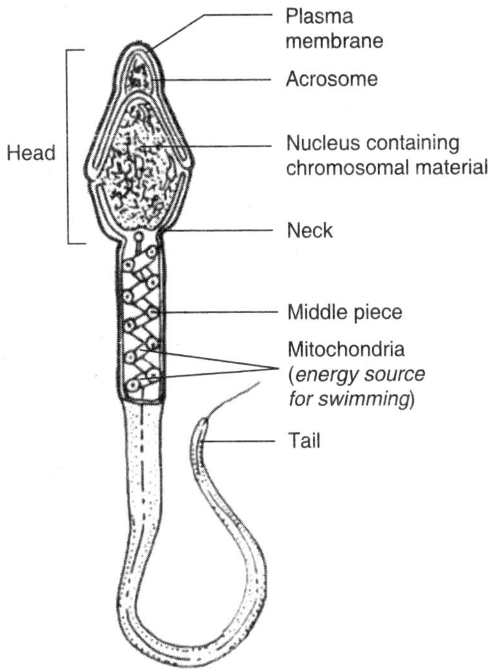

Fig. 9.14. Structure of a sperm.

(*ii*) *Neck* is very short and is present between the head and middle piece. It contains the **proximal centriole** towards the nucleus which plays a role in the first cleavage of the zygote and the **distal centriole** which gives rise to the axial filament of the sperm.

(*iii*) *Middle piece* possesses numerous mitochondria which produce energy for the movement of the sperm. At the end of the middle piece there is a **ring centriole (annulus)** with unknown function.

(*iv*) *Tail* is several times longer than the head. It consists of an axial filament surrounded by a thin layer of cytoplasm. The tail provide motality to the sperm, which is essential for fertilization.

The male ejaculates about 200 to 300 million sperms during a coitus. For a normal fertility, at least 60 per cent sperms of the ejaculate must have normal shape and size, and at least 40 per cent of the normal sperms must show vigorous motility. Sperms remain alive and retain their ability to fertilize an ovum from 24 to 48 hours after having been released in the female genital tract.

Oogenesis

The process of formation of a mature female gamete is called **oogenesis.** It takes place in ovaries and consists of three phases—multiplication, growth and maturation (Fig. 9.15).

(*i*) ***Multiplication Phase.*** During foetal development, certain cells in the germinal epithelium divide by mitosis producing a couple of million egg mother cells or oogonia in each foetal ovary. By the time the foetus is 25 weeks old, all the oogonia are already formed. These cells start division and enter into prophase I of the meiotic division and get temporily arrested at this stage, called **primary oocytes.**

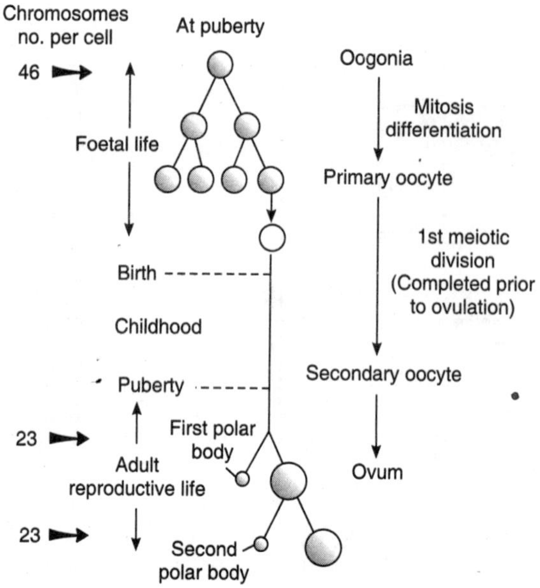

Fig. 9.15. Stages in oogenesis (diagrammatic).

(*ii*) ***Growth Phase.*** This phase begins when the female attains puberty, and may extend over many years. The primary oocyte enlarges and matures by taking food from the surrounding follicle cells. Usually, from two ovaries, only one primary oocyte matures at a time and once a month.

(*iii*) ***Maturation Phase.*** The mature primary oocyte undergoes its first meiotic division. It is an unequal division resulting in the formation of a large haploid **secondary oocyte** and a tiny **first polar body** or **polocyte.** The secondary oocyte retains bulk of the nutrient rich cytoplasm of the primary oocyte. It is not very certain whether the first polar body formed by the first meiotic division, divide further or degenerates. The secondary oocyte undergoes second meiotic division which does not proceed beyond metaphase until a sperm enters it. Ovulation occurs at this stage and the secondary oocyte is transferred to the fallopian tube. Entry of sperm into the secondary oocyte restarts the cell cycle breaking down MPF (M-phase Promoting Factor) and turning on APC (Anaphase Promoting Complex). Completion of meiosis II divides the secondary oocyte into a large ootid and a very small second polar body. The ootid grows into a functional ovum. Thus, from one oogonium, one ovum (female gamete) and three polar bodies are formed. The polar bodies soon degenerate.

Hormonal Control of Oogenesis. Oogenesis occurs under the influence of hormones, gonadotropin releasing hormone (GnRH) secreted by the hypothalmus , which stimulates the anterior pituitary to secrete Lutenising Hormone (LH) and Follicle Stimulating Hormone (FSH). FSH stimulates the growth of the ovarian follicles and division of primary oocyte within the follicle to form secondary oocyte. FSH also stimulates the corpus luteum to secrete progesterone. The rising level of progesterone inhibits the release of GnRH, which in turn inhibits the production of FSH, LH and progesterone (Fig. 9.16).

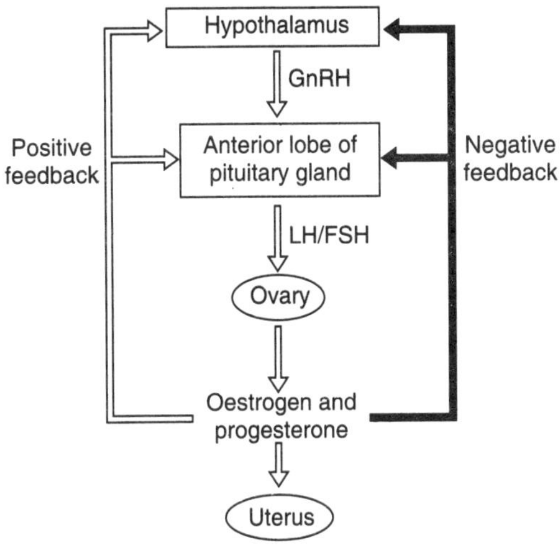

Fig. 9.16. Hormonal control of oogenesis.

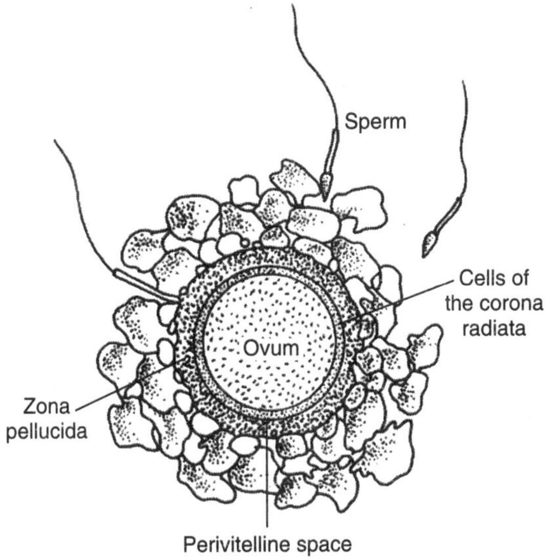

Fig. 9.17. Ovum surrounded by few sperms.

Structure of Ovum. The mature human ovum is spherical in shape and almost without yolk. Such an ovum is called **alecithal.** Its cytoplasm called **ooplasm,** contains large nucleus. The ovum is surrounded by a non-cellular transparent layer called **zona pellucida.** Outside this layer, there are radially arranged follicular cells forming anothor layer known as **corona radiata.** A narrow **vitelline space** is present between the zona pellucida and plasma membrane. The area of the ovum having polar bodies is called animal pole, while the area opposite to the animal pole is called **vegetal pole** (Fig. 9.17).

Differences between Spermatogenesis and Oogenesis

Spermatogenesis	*Oogenesis*
1. It occurs in the testis.	1. It occurs in the ovaries.
2. Spermatogonia change to primary spermatocytes.	2. Oogonia change to primary oocytes.
3. A primary spermatocyte divides to form two secondary spermatocytes.	3. A primary oocyte divides to form one secondary oocyte and one polar body.
4. The secondary spermatocyte divides to form two spermatids.	4. A secondary oocyte divides to form one ootid and one polar body.
5. A spermatogonium forms four spermatozoa.	5. An oogonium forms one ovum.
6. It is completed in the testis.	6. It begins in the ovary and is completed in the female reproductive tract.

Menarche

The beginning of menstruation or the first menstruation is called **menarche.** It starts at puberty at the age between 12 to 15 years.

MENSTRUAL CYCLE

The cyclic changes in the reproductive tract of the female primates (*e.g.,* monkeys, apes and human beings), associated with the release of ovum, and extends to more or less a month (menses) is called **menstrual cycle.** These changes end with a discharge of blood, unfertilized ovum and broken tissue of the uterine mucosa (the endometrium) through vagina, if pregnancy does not occur. This process is called **menstruation.** The menstruation is repeated at an average interval of about 28/29 days. One ovum is released (ovulation) during the middle of each menstrual cycle. The menstruation cycle is regulated by certain hormones secreted by the pituitary gland. The pituitary gland is stimulated by the releasing factors produced in the hypothalamus. The hormone produced by the pituitary gland influence the ovaries. The hormones secreted by the ovaries bring about changes in the uterine wall.

The menstrual cycle consists of three phases—menstrual phase, proliferative phase and secretory phase (Figs. 9.18 and 9.19).

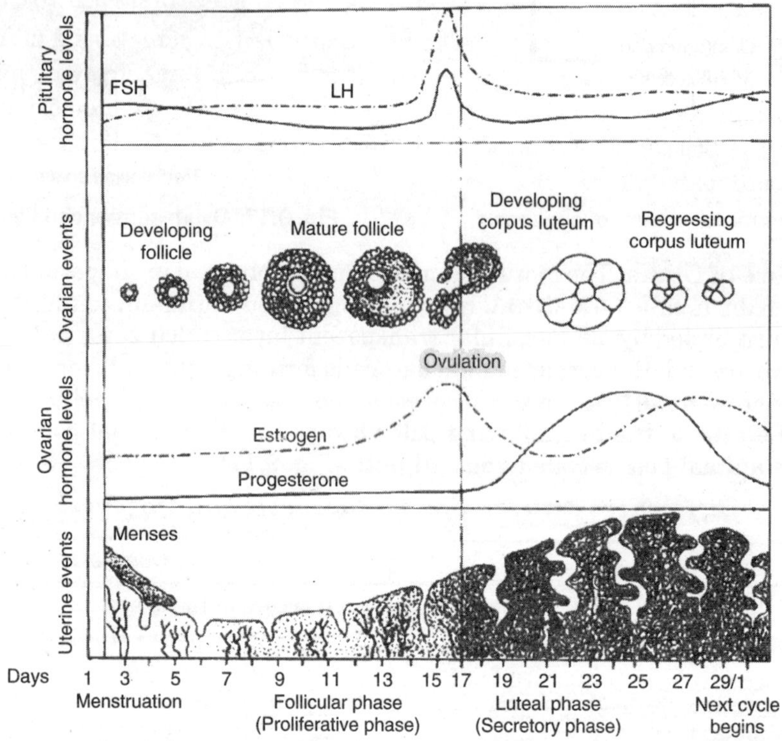

Fig. 9.18. Diagrammatic presentation of various events during a menstrual cycle.

1. Menstrual Phase (Bleeding phase or menses). The cycle starts with the menstrual phase, when there is menstrual flow. This phase lasts for about 3–5 days. The production of lutenising hormone (LH) from the anterior lobe of the pituitary gland is considerably reduced. The withdrawal of this hormone causes degeneration of the corpus luteum and therefore progesterone production is reduced. The endometrium of the uterus breaks down and menstruation begins. The broken endometrium, its secretions, blood and unfertilized ovum constitute the menstrual flow. Menstruation only occurs if the released ovum is not fertilized. Lack of menstruation may be indicative of pregnancy. However, it may also be caused due to some other reasons like stress, poor health, etc.

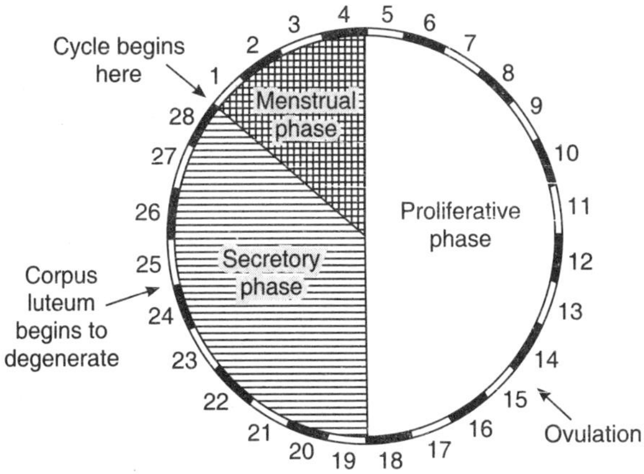

Fig. 9.19. Schematic representation of menstrual cycle.

2. Proliferative Phase (Follicular phase). The menstrual phase is followed by the follicular phase. It lasts for about 14 days in a 28 days cycle. The follicle stimulating hormone (FSH) secreted by the anterior lobe of the pituitary gland stimulates the ovarian follicles to secrete oestrogen. The primary follicle in the ovary grows to become a fully mature Graafian follicle. Estrogen stimulates the proliferation of endometrium of the uterine wall. The endometrium becomes thicker by rapid cell multiplication and this is accompanied by an increase of uterine glands and blood vessels. Both LH and FSH attain a peak level in the middle of cycle (about 14th day). Rapid secretion of LH leading to its maximum level during the mid-cycle called **LH surge.** It induces rupture of Graafian follicle and thereby the release of ovum (ovulation). At the same time the production of estrogen also stops.

3. Secretory Phase (Luteal phase). The ovulation (ovulatory phase) is followed by luteal phase, which lasts for about 10 days in a 28-day cycle. During this phase, the remaining parts of Graafian follicle transform as the **corpus luteum**. The corpus luteum secretes large amounts of **progestogens** (*e.g.*, **progesterone**). Progesterone stimulates the uterine glands to produce increased amounts of watery mucus. There is also similar increase in the secretion of watery mucus by the vaginal glands and by the glands of the Fallopian tubes.

Progesterone is also essential for the maintenance of the endometrium. Such as endometrium is necessary for implantation of the fertilized ovum and other events of pregnancy. During pregnancy all events of the menstrual cycle stop and there is no menstruation. In the absence of fertilization, the corpus luteum degenerates and changes into a white body called **corpus albicans.** This causes disintegration of the endometrium leading to menstruation, making a new cycle.

Menopause

In human beings, menstrual cycles ceases around 50 years of age. It is called **menopause.** Cyclic menstruation is an indicator of normal reproductive phase and extends between menarche and menopause. Some women have irregular cycles for months or years prior to menopause. Others simply stop menstruating abruptly. It is believed that menopause is the result of changes in pituitory gland and the nearby hypothalamus or the menopause occurs when no follicles are left in the ovaries. After menopause, there is reduced supply of estrogen and progesterone, because of the absence of developing follicles. As a result, women may suffer temporarily depression, hot flashes and other physiological and psychological problems.

FERTILIZATION

It is the process of fusion of the sperm with an ovum to form a diploid cell called **zygote.** During copulation (coitus), semen containing millions of sperms is deposited by the penis into the vagina. This process is called **insemination.** A single ejaculate of 3–4 ml of semen may contain 200 to 300 million sperms.

Once deposited within the vagina, the sperms proceed on their journey towards the fallopian tubes. The motile sperms swim rapidly, pass through the cervix, then enter into the uterus and finally reach the junction of the isthmus and ampulla (ampullary-isthmic junction) of the fallopian tube. Sperms move in the fluid medium at the rate of 1.5 to 3 mm per minute to reach the site. The leucocytes of the vaginal epithelium engulf millions of sperms, therefore only a few thousand find their way to reach the fallopian tubes.

The secondary oocyte (ovum) is released from the mature Graafian follicle of the ovary (ovulation). The oocyte is received by the nearby infundibulum and sent into the fallopian tube by the movements of fimbrae and their cilia. The released ovum is also transported to the ampullary-isthmic junction, where fertilization takes place. The secondary oocyte can be fertilized only within 24 hours after its release from the ovary. Fertilization can only occur if the secondary oocyte and sperms are transported simultaneously to the ampullary-isthmic junction. This is the reason, why not all copulations leads to fertilization and pregnancy. The secondary oocyte is surrounded by numerous sperms but only one sperm succeeds in fertilizing the oocyte. In secondary oocyte, second meiotic division is completed by the entry of the sperm into it. After this secondary oocyte is called ovum (Fig. 9.20).

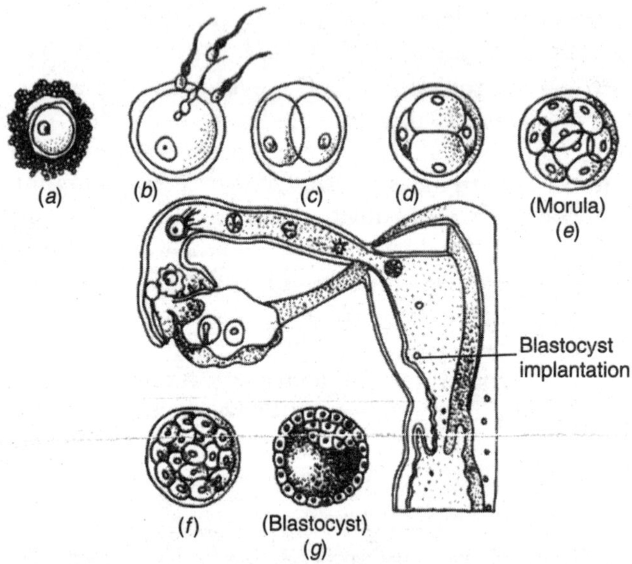

Fig. 9.20. Transport of ovum, fertilization and passage of growing embryo through fallopian tube.

Capacitation of Sperms. The secretions of the female genital tract make the sperms capable of fertilizing the ovum. These secretions remove the coating substances deposited on the surface of the sperm particularly those on the acrosome. Thus, the receptor sites on the acrosome are exposed and sperm becomes active to penetrate the egg. This phenomenon of sperm activation in mammals is known as **capacitation**. It takes about 5 to 6 hours for capacitation.

Physical and Chemical Events of Fertilization

In the process of fertilization, the sperm penetrates the secondary oocyte and enters its territory through intricate chemical reaction. This process is called acrosomal reaction (Fig. 9.21). The sperm releases chemical substances of enzymatic nature. These chemicals are collectively called **sperm lysins**. Important lysins are (*i*) **hyaluronidase** that acts on the ground substances of follicle cells, (*ii*) **corona penetrating enzyme** that dissolves corona radiata and (*iii*) **zona lysine** or **acrosin** that helps to digest the zona pellucida. Optimum pH, Ca^{2+}, Mg^{2+} ion concentration and temperature are essential for acrosomal reaction. Ca^{2+} ions plays a major role in acrosomal reaction. In absence of Ca^{2+}, fertilization does not occur.

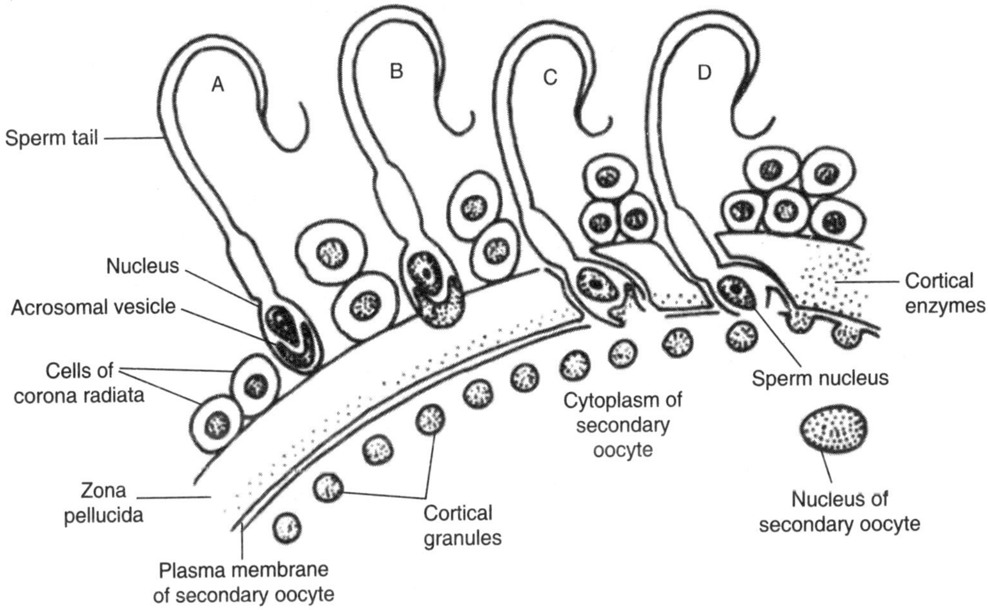

Fig. 9.21. Stages of sperm entry into the ovum during fertilization.

Due to acrosomal reaction, plasma membrane of the sperm fuses with the plasma membrane of the secondary oocyte so that the sperm contents enter the oocyte. Binding of the sperm to the secondary oocyte induces depolarization of the oocyte plasma membrane. Depolarization ensures **monospermy** (entry of one sperm into the oocyte) and prevents **polyspermy** (entry of more than one sperm into the oocyte). When the sperm fuses with the plasma membrane of the oocyte, the secondary oocyte shows a **cortical reaction.** The cortical granules present below the plasma membrane of the oocyte fuse with the latter and release their enzymatic content between the plasma membrane and the zona pellucida. The enzyme harden the zona pellucida which helps in prevention of the entry of additional sperms (polyspermy) into the secondary oocyte.

At the point of contact with the sperm, the secondary oocyte forms a projection called the **cone of reception** or **fertilization cone** which receives the sperm, the distal centriole of the sperm divides and forms two centrioles to generate the mitotic spindle formation for cell division. The mammalian secondary oocyte (ovum) does not have centrioles of its own.

The entry of sperm stimulates the secondary oocyte to complete the suspended second meiotic division. This produces a haploid mature ovum and a second polar body. The head of the sperm containing the nucleus, separates from the middle piece and tail, and becomes the

male pronucleus. The second polar body and the tail of the sperm degenerate. The nucleus of the ovum is now called **female pronucleus.** The male and female nuclei move towards each other and get fused. It is called **karyogamy** or **amphimixis**. The fertilized ovum is now called **zygote.** The zygote is a diploid cell that has 46 chromosomes.

Specificity in Fertilization. The adhesion of sperm to the egg of the same species occurs through a chemical interaction called **agglutination.** The ovum secretes a chemical called **fertilizin** (a glycoprotein = monosaccharides + amino acids) and the sperm possesses on its surface a protein called **antifertilizin** (composed of acidic amino acids). The fertilizin of an egg interacts with the antifertilizin of a sperm only of the same species.

Significance of Fertilization. *(i)* It restores the diploid number of chromosomes, characteristic of the species *(ii)* It combines the characters of two parents, thus introduces variation *(iii)* Fertilization introduces the centrioles which is lacking in the mature egg. *(iv)* It initiates cleavage.

Determination of Sex. In human beings the sex of the baby is determined at the time of fertilization. As you know, a human female has XX sex chromosomes and the male has XY sex chromosomes. Therefore, all the haploid gametes produced by the female (ova) have the sex chromosome X, whereas in the male gametes (sperms) the sex chromosome could be either X or Y. Thus, 50 per cent of sperms carry the X chromosome, while the other 50 per cent carry the Y chromosome. After fusion of the male and female gametes, the zygote would carry either XX or XY sex chromosomes depending on whether the sperm carrying X or Y chromosome fertilizes the ovum. The zygote carrying XX sex chromosome would be a female baby and XY sex chromosome would be a male baby. Thus, the sex of the child is determined by the father and not by the mother (Fig. 9.22).

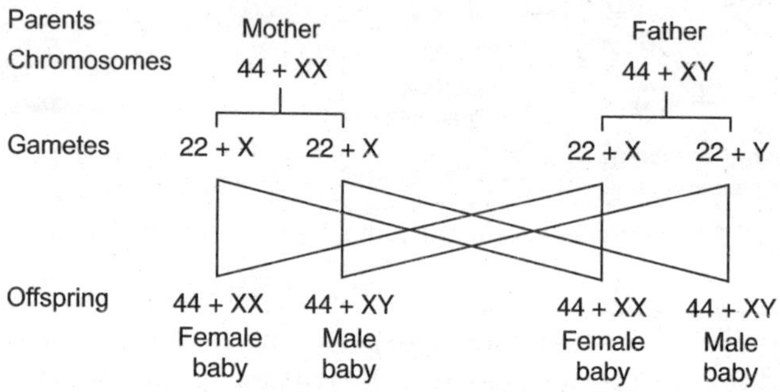

Fig. 9.22. Sex determination in human beings.

EMBRYONIC DEVELOPMENT

The development of embryo from fertilized ovum involves the following events.

Cleavage and Blastocyst Formation

About thirty hours after fertilization, the zygote (fertilized ovum), undergoes a series of mitotic divisions and forms (2, 4, 8, 16 daughter cells called **blastomeres.** The embryo with 8 to 16 blastomeres looks like a mulberry and is called **morula.** During cleavage the young embryo is slowly moving down the fallopian tube towards the uterus. The morula continues to divide and transforms into a sphere of cells with a cavity is the centre. This cavity is called **blastocoel** and the embryo is termed the **blastocyst.** The blastocyst consists of an outer envelope of cells, the **trophoblast** or **trophoectoderm** and **inner cell mass** (= embryoblast).

The side of the blastocyst to which the inner cell mass is attached is called the **embryonic** or **animal pole,** while the opposite side is the **abembryonic pole.** The inner cell mass of the blastocyst is the precursor of the embryo *i.e.*, gives rise to embryo. The cells of trophoblast help to provide nutrition to the embryo. They later form the extra embryonic membranes namely chorion and amnion and part of the placenta. The cells of the trophoblast, which are in contact with the inner cell mass are called **cells of Rauber.**

Implantation

The attachment of the blastocyst to the uterine wall is called **implantation.** It occurs after 7 days of fertilization. The blastocyst sinks into the endometrium of the uterus from the side where the inner cell mass is located, and gets completely burried in the endometrium. It establishes the **pregnancy** (Fig. 9.23). Pregnancy refers to the time from conception to birth. In human beings it is appoximately 9 months ± 7 days. The duration of pregnancy in dogs, elephants and cats is 63, 624 and 63 days respectively. Zona pellucida layer which is still present over blastocyst plays a significant role in implantation. It does not expose the sticky and phagocytic cells of the trophoblast till the blastocyst reaches the proper implantation site. When the blastocyst reaches the proper site, zona pellucida becomes thinner and finally disappears.

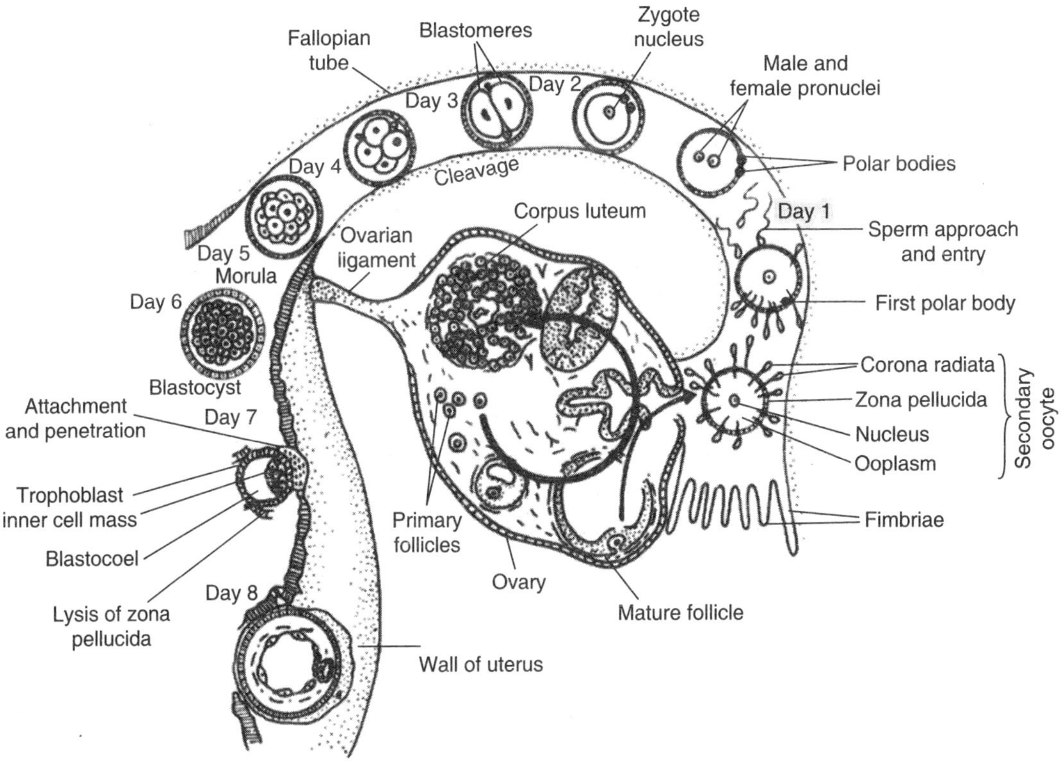

Fig. 9.23. Diagram showing ovulation, fertilization, cleavage and implantation of blastocyst.

Formation of Placenta and Decidua

After implantation, finger like projections appear on the trophoblast called **chorionic villi,** which are surrounded by the uterine tissue and maternal blood. The chorionic villi and uterine tissue become interdigited with each other and jointly form a structural and functional unit between developing embryo (foetus) and maternal body called **placenta.**

After implantation, the uterine mucosa (endometrium) undergoes profound changes to form **decidua.** The decidua is shed when the foetus is delivered. The decidua tissue is differentiated into three layers.

(*i*) ***Decidua Basalis.*** It is the portion of the endometrium between the chorion and the myometrium of the uterus. The decidua basalis becomes the maternal part of the placenta.

(*ii*) ***Decidua Capsularis.*** It is the portion of the endometrium between the embryo and uterine cavity and

(*iii*) ***Decidua Parietalis.*** It is the portion of the modified endometrium that lines the entire pregnant uterus, except for the area where the placenta is forming.

The developing embryo obtains nutrients and oxygen and gets rid of carbon dioxide and metabolic wastes through placenta. The growing embryo develops four extra embryonic or foetal membranes : **amnion, chorion, allantois** and **yolk sac.** Amnion provides a fluid medium to the developing embryo, prevent desiccation and also act as shock absorber. The chorion and allantois take part in the formation of placenta (*chorioallantoic placenta e.g.,* most eutherian mammals). In humans placenta develops from chorion. Because the chorion take part in the formation of placenta, the human placenta is called the **chorionic placenta.** It consists of foetal part, the chorion and a maternal part the decidua basalis. The foetal part of placenta grows to invade the uterine mucosa with its chorionic villi. The degree of intimacy is so strong that the blood vessels of the chorionic villi are bathed in the mother's blood. This is due to erosion of the uterine mucosa, including its epithelium, connective tissue and the endothelial lining. This type of placenta which is based on the intimacy between foetal and maternal parts of the placenta, is termed as **haemochorial placenta.** The placenta is connected to embryo through an **umbilical cord** which helps in the transport to substances to and from the embryo (Fig. 9.24).

Functions of Placenta. (*i*) Placenta facilitates the supply of oxygen and nutrients from maternal blood to the embryo (*ii*) The carbon dioxide and excretory/waste materials produced by the embryo also pass to maternal blood through placenta and are excreted by the mother. (*iii*) Placenta acts as an ultrafilter and allow to pass into foetal blood, the nutrients, hormones and antibodies against diseases like diphtheria, small pox, scarlet fever, measles etc. (*iv*) Placenta also acts as an endocrine tissue and produces several hormones like **human chorionic gonadotropin (hCG), human placental lactogen (hPL), estrogens, progestogens** etc. The hCG stimulates the corpus luteum that continues to secrete progesterone until the end of pregnancy. (*v*) In the later phase of pregnancy, it secretes—**relaxin** that facilitates parturition (act of birth) by softening the connective tissue of the pubic symphysis. Relaxin is also secreted by the ovary. It is important to note that hCG, hPL and relaxin are produced in women only during pregnancy.

In addition, the levels of other hormones like oestrogens, progestogens, cortisol, prolactin, thyroxine, etc. are increased several folds in the maternal blood. Increased production of these hormones is essential for supporting the fetal growth, metabolic changes in the mother and maintenance of pregnancy.

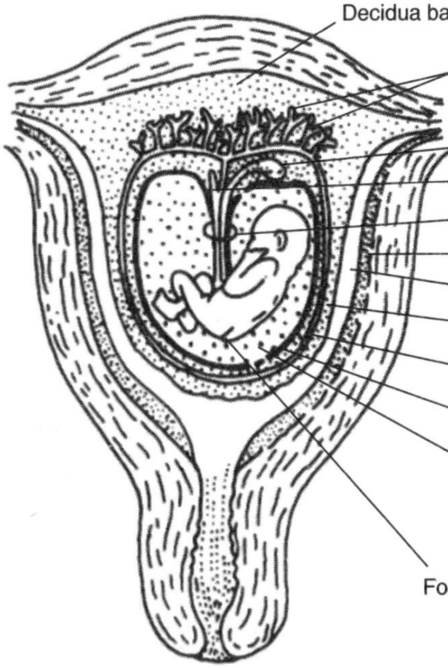

Fig. 9.24. Human foetus within the uterus showing placenta.

Role of Human Chorionic Gonadotropins (hCG). The human chorionic gonadotropin hormone is secreted by trophoblastic cells of the blastocyst. It has properties similar to those of Luteinizing Hormone (LH) of the pituitary gland. It takes over the job of pituitary LH during pregnancy. The hCG maintains the corpus luteum and stimulate it to secrete progesterone. The latter maintains the endometrium of the uterus and causes it to grow throughout the pregnancy. This also prevents menstruation. Progesterone also increases the secretion of mucus in the cervix of the uterus to form a protective plug during pregnancy.

Formation of Germ Layers (Gastrulation)

As you have read that early blastocyst consists of inner cell mass and trophoblast. The inner cell mass contains cells called **stem cells,** which have the potency to give rise to all tissues and organs. The cells of inner cell mass of blastocyst move, that help to attain new shape and morphology of the embryo. These cell movements are called **morphogenetic movements.** The inner cell mass differentiates into outer layer called **ectoderm** and an inner layer called **endoderm**. A **mesoderm** soon appears between the ectoderm and the endoderm. The morphogenetic movements transforms the blastocyst (blastula) into a three layered structure called **gastrula.** The transformation of the blastocyst into the gastrula with primary germ layers by rearrangement of the cells is called **gastrulation.** The three primary germ layers of the gastrula give rise to all tissues and organs in adults. The various tissues and organs derived from primary germ layers are given in Table 9.1.

Table 9.1. Tissues and Organs derived from Primary Germ Layers

Ectoderm	*Mesoderm*	*Endoderm*
1. Epidermis and epidermal glands of skin.	1. Dermis of skin.	1. Mucus membrane of gut.
2. Epithelium of stomodeum and proctodaeum.	2. All muscular tissues.	2. All digestive glands.
3. Enamel of teeth.	3. All connective tissue.	3. Thyroid, parathyroid and thymus glands.
4. Entire nervous system.	4. Coelom, coelomic epithelium and mesenteries.	4. Middle ear and eustachian tube.
5. Conjuctiva, lens, cornea and retina of eyes.	5. Dentine of teeth.	5. Trachea, bronchi and lungs.
6. Nasal and olfactory epithelia.	6. Pericardium and pleura.	6. Urinary bladder.
7. Auditory vesicles and internal ear.	7. Notochord and most skeleton.	—
8. Pituitary and pineal glands and adrenal medulla.	8. Heart and blood vessels.	—
9. Cartilages and visceral arches.	9. Spleen.	—
	10. Excretory system.	—
	11. Reproductive system.	—
	12. Sclerotic and choroid of eyes.	—
	13. Blood and lymph.	
	14. Adrenal cortex.	

Changes in Embryo. In human beings, the heart of embryo is formed after one month of pregnancy. The heart sound of the foetus can be detected through the stethoscope. By the end of second month the foetus develops limbs and digits. By the end of first trimester (about 12 weeks) most of the major organs systems (*e.g.*, the limbs and external genitalia) are well developed. The basic structure of the foetus is formed during the first trimester (first 3 months).

During this period of foetus is very sensitive to certain foreign agents. Virus infection of the mother *e.g.*, rubella virus (German measles) or exposure to certain chemicals may cause malformations in the developing foetus. Such agents are called **teratogens** (monster forming). The most well known synthetic teratogen drug is **thalidomide.** The drug causes multiple defects in the developing foetus.

The first movement of the foetus and hair are generally observed during the fifth month. By the end of second trimester (about 24 weeks) the hair appear on the entire body, eye lids separate and eye lashes are developed. By the end of 9 months ± 7 days of pregnancy, the foetus is fully formed and is ready for delivery.

PARTURITION AND LACTATION

Parturition

The process of delivery of the foetus (child birth) is called parturition. The duration of pregnancy in human beings is about 9 months ± 7 days. At the end of pregnancy, vigorous contraction of uterus causes delivery or expulsion of the foetus. Parturition is induced by both nervous system and hormones secreted by the endocrine glands of mother. The signals for child birth (parturition) originate from the fully developed foetus and placenta which induce mild uterine contraction called **foetal ejection reflex.** This triggers quick release of **oxytocin** from the maternal posterior lobe of pituitary gland. Oxytocin acts on the uterine muscle and causes stronger uterine contractions, which in turn stimulates further secretion of oxytocin. The stimulatory reflex between the uterine contraction and oxytocin secretion continues resulting in stronger and stronger contractions. Hormone **relaxin** released by the placenta and ovary, increases the flexibility of the pubis symphysis and ligaments of the sacroiliac and sacrococcygeal joints and helps dilate the uterine cervix during **labour pains** (pains of child birth). Both of these actions give relief to the body from the pain during delivery of the baby. With continued forceful uterine contractions, the amnion ruptures and the amniotic fluid (the "waters") flows out, lubricating the vagina. This leads to expulsion of the baby out of the uterus through the birth canal. Soon after the infant is delivered, the placenta, umblical cord and foetal membranes are expelled. These structures are referred as **after birth** or **decidua.** Recently a hormone called **corticotropin releasing hormone** (CRH), which in non-pregnant women is secreted only by neuro-secretary cells of hypothalamus, is produced by the placenta in pregnant women. CRH is now thought to be a part of the 'clock' that establishes the timing of birth. Secretion of CRH by placenta increases enormously towards the end of pregnancy. Women who have higher levels of CRH earlier in pregnancy are more likely to deliver prematurely, whereas those have low levels are more likely to deliver after their due date.

Lactation

The mammary glands of the female undergo differentiation during pregnancy and starts producing milk towards the end of pregnancy and after the birth of the baby, by the process called **lactation.**

At puberty in females, mammary glands begin to develop under the influence of estrogen and progesterone. Secretion and storage of milk generally begins, usually within 24 hours, after the birth of young one, under the influence of hormone **prolactin** secreted by anterior lobe of pituitary gland. However, the ejection of milk is stimulated by the hormone **oxytocin** released from the posterior lobe of the pituitary gland. The milk produced during the initial few days of lactation is called **colostrum**. This is yellowish fluid, which is rich in protein (lactalbumin and lactoprotein) but low in fatcolostrum contains several antibodies (I_gA is the major immunoglobulin), that provide passive immunity to the new born baby. Therefore, breast feeding during the initial period of infant growth is recommended by doctors for bringing up a healthy baby. Milk production in the mammary gland is under feedback control. Milk contains an **inhibitory peptide.** If the mammary glands are not fully emptied, the peptide accumulates and inhibits milk production.

REVIEW EXERCISES

MULTIPLE CHOICE TYPE QUESTIONS

1. What is the site of fertilisation in human beings?
   (*a*) Vagina (*b*) Uterus
   (*c*) Fallopian tube (*d*) Ovary.
2. Sertoli cells are found in
   (*a*) Seminiferous tubules and provide nourishment to germ cells
   (*b*) Ovaries and secrete progesterone

(*c*) Interstitial space and produce androgens
(*d*) Uterus and produce ovum.

3. How many sperms and ova will be produced from 25 primary spermatocytes and 25 primary oocytes respectively ?
(*a*) 100 sperms and 50 ova
(*b*) 100 sperms and 25 ova
(*c*) 50 sperms and 25 ova
(*d*) 100 sperms and 100 ova.

4. Acrosome reaction in sperms is triggered by
(*a*) Release of fertilizin
(*b*) Capacitation
(*c*) Release of lysin
(*d*) Influx of Na^+ into the sperm.

5. During copulation a slimy alkaline fluid to lubricate vagina is secreted by
(*a*) Perineal gland
(*b*) Cower's gland
(*c*) Prostrate gland
(*d*) None of these.

6. The part of Fallopian tube closest to the ovary is
(*a*) Cervix
(*b*) Ampulla
(*c*) Isthmus
(*d*) Infundibulum.

7. Foetal ejection in human females is induced by
(*a*) Pressure exerted by ammiotic fluid
(*b*) Release of oxytocin from pituitary
(*c*) Fully developed foetus and placenta
(*d*) Differentiation mammary glands.

8. For parturition oxytocin is released from
(*a*) Ovaries
(*b*) Uterus
(*c*) Placenta
(*d*) Pituitary gland.

9. Corpus luteum releases
(*a*) Estrogen
(*b*) Progesterone
(*c*) Estrogen and progestone
(*d*) Androgen.

10. About which day in a normal human menstrual cycle does rapid secretion of LH (popularly called LH surge) normally occur ?
(*a*) 5th day
(*b*) 11th day
(*c*) 14th day
(*d*) 20th day.

11. The first movement of the foetus and appearance of hair on its head are usually observed during which month of pregnancy ?
(*a*) Sixth month
(*b*) Third month
(*c*) Fourth month
(*d*) Fifth month.

12. Acrosome of a sperm is a modified
(*a*) Lysosome
(*b*) Golgi complex
(*c*) Nucleus
(*d*) Mitochondria.

OBJECTIVE TYPE QUESTIONS

1. Fill in the blanks with suitable words :
(*i*) Testes are housed in a sac named ....
(*ii*) Corpus luteum secretes .... while leydig's cells secrete ....
(*iii*) An oogonium produces one .... and .... polar bodies.
(*iv*) If ovum is not fertilised then the corpus luteum degenerates into a white mass called ....
(*v*) The part of uterus which projects into vagina is known as ....
(*vi*) Stoppage of egg formation and menstrual cycle in a woman is called ....
(*vii*) Urethra in the male carries both urine and ....

2. Mark the following statements as True (T) or False (F) :
   (*i*) The first menstruation begin at puberty is called menopause.
   (*ii*) Sertoli cells provide nourishment to the developing sperms.
   (*iii*) In man seminal vesicle is a paired gland, whereas prostate and Cowper's are unpaired glands.
   (*iv*) During fertilisation the entire sperm enters into the egg.
   (*v*) In the virgin, the vaginal orifice is covered by a membrane called hymen.
3. Match the terms of Column I with appropriate terms of Colunm II.

Column I	**Column II**
(*i*) Leydig's cells	(*a*) Spermatozoa
(*ii*) Spermiogenesis	(*b*) Progesterone
(*iii*) Endometrium	(*c*) Testosterone
(*iv*) Acrosome	(*d*) Spermatids
(*v*) Corpus luteum	(*e*) Uterus

TEST QUESTIONS

1. Why do testes descend in the scrotum ?
2. What is semen ?
3. What stimulates pituitary to release the hormone responsible for parturition ? Name the hormone and its source organ.
4. Mention the functions of the following :
   (*i*) Fallopian tubes
   (*ii*) Uterus
   (*iii*) Seminiferous tubules
   (*iv*) Graafian follicle.
5. How is placenta formed in human female ? Write its functions.
6. What is corpus luteum ? What is its function ? Under what condition it undergoes degeneration.
7. Differential between menarchae and menopause.
8. Describe the formation blastocyst in a human embryo.
9. What is spermatogenesis ? Describe the process of spermatogenesis.
10. Describe the structure of human female reproductive system.
11. Explain the process of oogenesis in a human female.
12. With the help of a diagram, describe the structure of a human sperm.
13. Give an account of menstrual cycle in human female.
14. Write the specific location and the functions of the following cells in human males :
    (*i*) Leydig cells
    (*ii*) Sertoli cells
    (*iii*) Primary spermatocytes.
15. Explain the process of parturition in human female.

10 Reproductive Health

- Reproductive Health—Problems and Strategies
- Population Explosion and Birth Control
  - — Reasons for Population Explosion
  - — Measures to Control Overpopulation
- Medical Termination of Pregnancy (MTP)
- Sexually Transmitted Diseases (STDs)
- Infertility
- Review Exercises
- Test Questions

Reproduction is the fundamental property of a living being and is essential for multiplication and perpetuation of a species. Normal reproduction is related with normal reproductive health. The term 'reproductive health' refers *to healthy reproductive organs with normal functions*. According to the World Health Organisation (WHO), reproductive health means a total well being in all aspects of reproduction *i.e.,* physical, emotional, behavioural and social. Thus, '*the persons having physically and functionally normal reproductive organs and normal emotional and behavioural interactions among them in all sex related aspects may be called reproductively healthy*'. Therefore, for a healthy society, it is important to maintain reproductive health. Let us discuss the methods taken up to achieve the reproductive health.

REPRODUCTIVE HEALTH—PROBLEMS AND STRATEGIES

1. Overpopulation. India is a populous country. The large population is the main problem of our country, which is also concerned with reproductive health. India was amongst the first country in the world to initiate action plans and programmes at a national level to attain total reproductive health as a social goal. One such programme is 'Family Planning Programme', which was initiated in 1951 and was periodically assessed. Presently, improved programmes covering wider reproduction related areas are in operation under the popular name **'Reproductive and Child Health Care (RCH) Programmes'.** The major objectives of these programmes are to create awareness among people about various aspects related to reproduction and to provide facilities and support for building up a reproductively healthy society.

2. Awareness about Reproduction. Government and Non-Government Agencies (NGOs) have taken various steps to create awareness among the people about reproduction-related aspects with the help of audio-visual and the print media. Parents, other close relatives, teachers and friends also have a major role in giving information about the reproduction related aspects.

3. Sex Education. Introduction of sex education in school should be encouraged so as to provide right information to the youth. Such education will also discourage children from believing in myths and having misconceptions about sex related aspects.

4. Adolescence Related Changes and STDs. It is important to provide proper information to the youngs about reproductive organs, adolescence (period of rapid growth between childhood and adulthood), safe and hygienic sexual practices, Sexually Transmitted Diseases (STDs), AIDS etc. It would help to lead a reproductive healthy life.

5. Birth Control and Care of Mother and Child. The target of socially conscious healthy families of desired size could be achieved by educating people, especially fertile couples and those in marriageable age-group, about birth control options, care of pregnant mothers, post-natal care of the mother and child, importance of breast feeding, equal opportunities for the male and the female child, etc.

6. Sex Abuse and Sex Related Crimes. Awareness of problems due to uncontrolled population growth, social evils like sex-abuse and sex related crimes etc. need to be created to enable people to think and take up necessary steps to prevent them and thereby build up a socially responsible and healthy society.

7. Infrastructural Facilities and Material Support. To attain reproductive health of the society various action plans such as strong infrastructural facilities, professional expertise and material support should be successfully implemented. These are essential to provide medical assistance and care to people in reproduction-related problems like pregnancy, delivery, STDs, abortions, contraception, menstrual problems, infertility, etc. Implementation of better techniques and new strategies from time to time are also required to provide more efficient care and assistance to people. There must be statuatory ban on **Amniocentesis** (a foetal sex determination test based on chromosomal pattern in the amniotic fluid surrounding the developing embryo) for sex determination to legally check increasing female foeticides. There must be massive child immunisation programme to achieve the target of reproductive health.

Amniocentesis. It is a technique of finding out disorders (if any) and sex of the foetus based on the chromosomal pattern in the amniotic fluid surrounding the developing embryo. In this technique, a sample of amniotic fluid from the womb of a pregnant woman is taken out, using a special surgical syringe and the chromosomes of the foetal cells or amniocytes (cells of amniotic membrane and foetal skin) present in the amniotic fluid are analysed (Fig. 10.1). In this technique, the cells present in the amniotic fluid are cultured and stimulated to divide. The dividing cells are then stained and observed under the microscope. The chromosomes of the cells are counted and compared with those of normal human chromosomes for detecting any possible chromosomal abnormalities in the developing foetus.

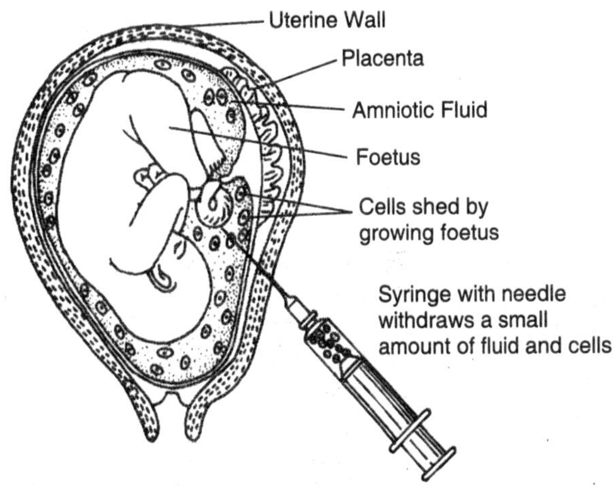

Fig. 10.1. Amniocentesis.

The amniocentesis technique has been developed for detecting foetal abnormalities by analysing the chromosal defect, if any of the foetus. But this technique is being misused for knowing the sex of the foetus, and to kill the normal female foetus. Amniocentesis is legally banned for the determination of sex to avoid female foeticide.

8. Research on Reproduction Related Areas. Research on various reproduction related areas are encouraged and supported by the governmental and non-governmental agencies to find out new methods and/or to improve upon the existing ones. For instance **Saheli,** a new oral contraceptive for the females was developed by the scientists at **Central Drug Research Institute (CDRI)** in **Lucknow, India.**

9. Medical Facilities. Better awareness about sex-related matters and the overall increased medical facilities for all sex-related problems indicate improved reproductive health of the society. These medical facilities must aim at increased number of medically assisted deliveries and better post-natal care leading to decreased maternal and infant mortality rates, increased number of couples with small families, better detection and cure of STDs etc.

POPULATION EXPLOSION AND BIRTH CONTROL

Population is defined as *'the total number of individuals of a species present in a particular area at a given time.'* A species has many populations living in different regions. The scientific study of human population is called **demography.** It involves three major aspects ; (*i*) changes in population size (increase or decrease), (*ii*) the composition of population and, (*iii*) the distribution of population. An official counting of population and preparing data about age-groups, births, deaths, sex ratio education etc., is called **census.** It is conducted as per the provision made under the census Act, 1948. In India, first census was carried out in 1891, and since then, it has been conducted regularly at interval of ten years.

In the last century, there has been significant improvement in the quality of life of the people, due to an overall development in various fields. However, increased health facilities along with better living conditions had an explosive impact on the growth of population. The world population which was around 2 billion (2000 million) in 1900, reached to about 6 billions by 2000. A similar trend was observed in India too. Our population which was approximately 350 million at the time of our independence. It reached close to the billion mark by 2000 and crossed 1 billion in May 2000. It means every sixth person in the world is an Indian. The rapid increase in human population over a relatively short period is called **population explosion.** The increase in human population with such an alarming rate could lead to an absolute scarcity of even basic needs, *i.e.*, food, shelter and clothing inspite of significant progress made in these areas during the past few decades.

Reasons for Population Explosion

The human population is largely due to following reasons:

(*i*) A rapid decline in death rate, Maternal Mortality Rate (MMR) and Infant Mortality Rate (IMR), due to better medical facilities.

(*ii*) An increase in number of people in reproducible age.

(*iii*) Better public health care and greater medical attention.

(*iv*) Control of certain diseases like plague, smallpox etc.

(*v*) Protection from natural calamities.

(*vi*) Better availability of food due to advancement in agriculture and improvement of storage and transport facilities.

(*vii*) Opposition of family planning programme by certain religions.

Though, our RCH (Reproductive and Child Health Care) programmes could bring down the population growth rate, but was only marginal. According to the census report of 2001, the population growth rate was still around 1.7 per cent *i.e.,* 17/1000/year. *i.e.,* a rate at which our population could double in 33 years.

Measures to Control Overpopulation

1. Education. For educating people to accept a small family norm the government has to create awareness about population explosion and to motivate people to have small families. People, particularly those in the reproductive age-group, should be educated about the advantage of small family. Non-governmental agencies (NGOs), mass media and educational institutions can play significant role in this compaign. For instance, the advertisements in the media and posters showing a happy couple with two children with a slogan '**Hum Do Hamare Do**' (we two our two) is one of the example of this compaign. Many couples, mostly the young, urban working ones have even adopted one child norm.

2. Raising of Marriageable Age. Statutory raising of marriageable age is one of the effective means to control the population. Presently the marriageable age of female is 18 years and that of male is 21 years.

3. Incentives. Certain incentives can be given to the couples with small families to tackle the problem of overpopulation. These incentives may be in the form of promotion in the service of the couples adopting one child norm or free education of the child.

4. Family Planning. The couples of reproductive age are motivated to plan their families by adopting birth control measures.

Birth Control

Birth control is the prevention of birth of a child by using some kinds of contraceptive measures.

Ideal Contraceptive. An ideal contraceptive should be (*i*) user friendly, (*ii*) easily available and reversible, (*iii*) with no or least side effects, and (*iv*) should not interfere with the sexual drive, desire and/or the sexual act of the user.

The commonly used contraceptive methods which help prevent unwanted pregnancies are described below. These are broadly grouped into natural/traditional, barrier, IUDs, oral contraceptives, injectables, implants and surgical methods.

1. Natural Methods. These methods work on the principle of avoiding chances of meeting of ovum and sperms. The various natural contraceptive methods are:

(*i*) ***Periodic abstinence (Avoidance of sex).*** It is one such method in which the couples avoid or abstain from coitus from day 10 to 17 of the menstrual cycle when ovulation could be expected. This period is called the **fertile period,** and the chances of fertilization are very high during this period. Therefore, the conception could be prevented by abstaining from coitus during the fertile period.

(*ii*) ***Withdrawal or Coitus Interruptus.*** This method needs good understanding between the wife and husband. The male partner withdraws his penis from the vagina just before ejaculation so as to avoid insemination.

(*iii*) ***Lactational Amenorrhea.*** (*Absence of menstruation*). This method is based on the fact that the menstrual cycle and therefore, ovulation do not occur during intense lactation after parturition. Therefore, as long as the mother breast feeds the child fully, chances of conception are almost nil. However, this method is effective only up to maximum period of six months after child birth.

2. Barrier Methods. In these methods ovum and sperms are prevented from physically meeting with the help of barriers. These barriers are available for both males and females.

(*i*) ***Condoms.*** These are barriers made of thin rubber/latex sheath that are used to cover the penis in the male or vagina and cervix in the female (Fig. 10.2), just before coitus so that the ejaculated semen would not enter the female reproductive tract. Condoms prevent conception and also protect the user from contracting STDs and AIDS. Condoms are available for both male and female, which can be self-inserted and thereby give privacy to the user. Such condoms are disposable. A popular brand of condom for male is **'Nirodh'.** The use of condoms has increased in the recent years, as they provide protection against sexually transmitted diseases.

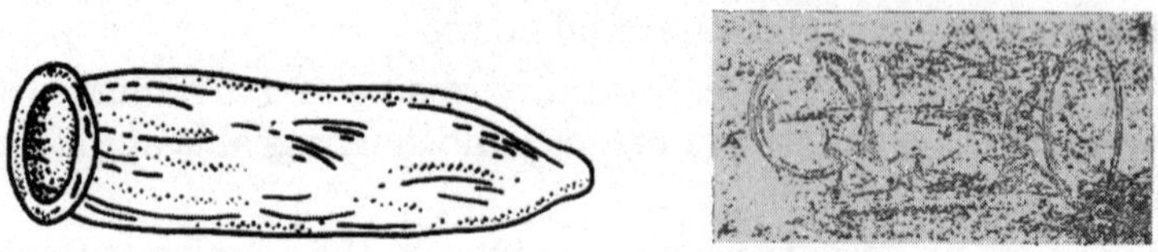

Fig. 10.2. A. Condom for male. B. Condom for female.

(*ii*) ***Diaphrams, Cervical Caps and Vaults.*** These barriers are made of rubber and are inserted into female reproductive tract to cover the cervix during coitus. They block the entry of sperms through the cervix, and thus prevent conception. These barriers are reusable and are used only at the time of coitus. Spermicidal creams, jellies and foams are generally used along with these barriers to increase their efficiency.

3. Intra-Uterine Devices (IUDs). These are plastic devices which are available in various shapes, and are inserted in the uterus through vagina by doctors or expert nurses. Presently, Intra-Uterine Devices are available in following three forms:

(*i*) ***Non-medicated IUDs*** (*e.g.*, Lippes loop). These increase phagocytosis of sperms within the uterus.

(*ii*) ***Copper releasing IUDs*** (*e.g.*, Cu T, Cu 7, Multi load 375). These increase phagocytosis of sperms. The Cu ions released by such IUDs (Fig. 10.3) suppress sperm motility and the fertilizing capacity of sperms.

(*iii*) ***Hormone releasing IUDs.*** These IUDs increase phagocytosis of sperms and also make the uterus unsuitable for implantation and the cervix hostile to the sperms, thus prevent pregnancy.

IUDs are ideal contraceptives for the females, who want to delay pregnancy and/or space between their children. In India, IUDs are most widely accepted method of contraception.

Fig. 10.3. Copper T (Cu T).

4. Oral Contraceptive Pills. These are hormonal tablets taken orally and hence are commonly called 'pills'. Small doses of either progesterones or progesterone-estrogen combination are used by the females (*e.g.*, Mala-D, and Mala-N). The contraceptive pills have to be taken daily for a period of 21 days starting preferably within the first five days of menstrual cycle. After a gap of 7 days (period of menstruation) it has to be repeated in the same pattern till the female-desires to prevent conception. Hormonal pills prevent ovulation and implantation by inhibiting the secretion of FSH and LH from the pituitary. They also alter the quality of cervical mucus to prevent/retard entry of sperm. Pills are very effective with lesser side effects and are well accepted by the females. A new oral contraceptive for the female named **'Saheli'** contains a non-steroidal preparation. It is a **once a week** (taken once in a week) pill with very few side effects and high contraceptive value.

5. Injections and Implants. The hormones like progestogens alone or in combination with estrogen are used by the females as injections or implants (Fig. 10.4) under the skin. Their action is almost similar to that of pills, but their effective periods are much longer.

The administration of progestrogens or progestogen-estrogen combinations or IUDs within 72 hours of intercourse is very effective as emergency contraceptive. They can be used to avoid pregnancy due to rape or casual unprotected intercourse.

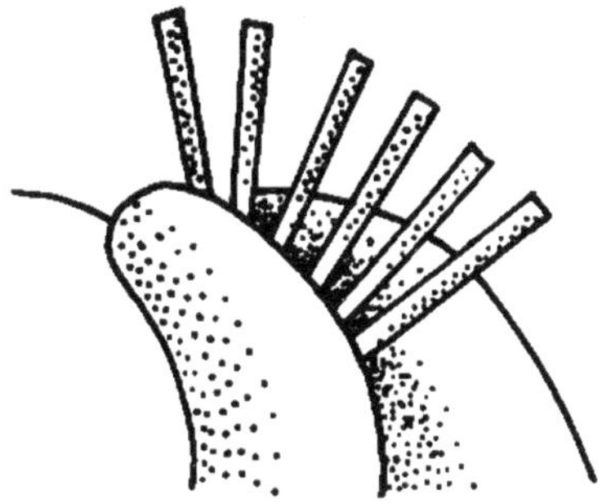

Fig. 10.4. Implants.

6. Surgical Methods. These methods involve **'sterilisation'** of male/female partner, and are used as terminal methods to prevent any more pregnancies. Surgical intervention blocks gamete transport and thereby prevent conception.

(*i*) ***Vasectomy.*** The sterilisation procedure in the male is called vasectomy. In vasectomy, a small part of the vas deferens is removed or tied up through a small incision on the scrotum (Fig. 10.5).

(*ii*) ***Tubectomy.*** The sterilisation procedure in the female is called tubectomy. In tubectomy, a small part of the fallopian tube is removed or tied up through a small incision in the abdomen or through vagina (Fig. 10.6).

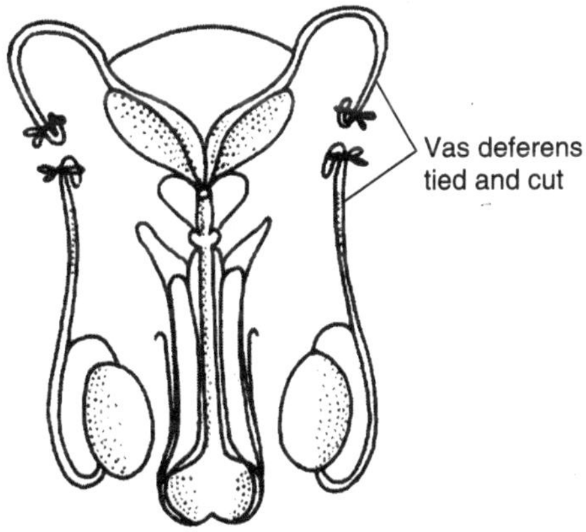

Fig. 10.5. Vasectomy.

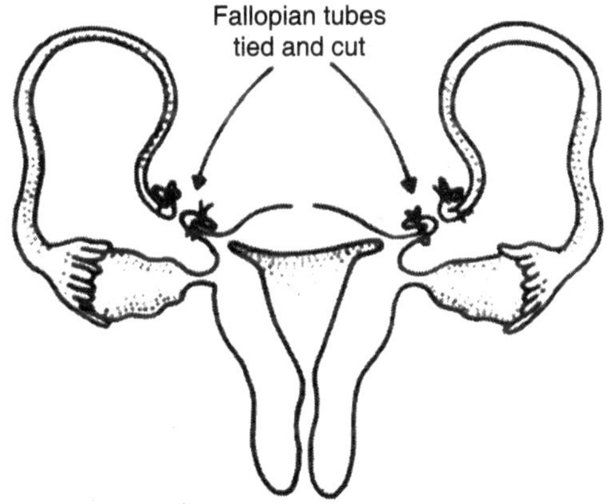

Fig. 10.6. Tubectomy.

Both vasectomy and tubectomy are very effective but their reversibility is very poor.

It is important to say here that the selection of a suitable contraceptive method and its use should always be undertaken in consultation with qualified doctors. One must also remember that the contraceptives are against natural conception/pregnancy. They are not also regular requirements for maintaining good reproductive health. Inspite of that one has to forced to use these methods to prevent pregnancy or to delay or space pregnancy. Although the widespread use of various contraceptive methods have significant role in checking uncontrolled growth of population, their possible ill effects (not very significant) should not be totally ignored. These include nausea (an inclination to vomit), abdominal pain, breakthrough bleeding, irregular menstrual bleeding or even breast cancer.

MEDICAL TERMINATION OF PREGNANCY (MTP)

Medical termination of pregnancy refers to voluntary or intentional termination of pregnancy before full term. It is also called **induced abortion.** About 45 to 50 millions MTPs are performed in a year in the world which is 1/5th of the total number of pregnancies in a year. Thus, MTPs help significantly in controlling human population. Although, it is not meant for that purpose. It is a debatable question in many countries, whether to accept/legalise MTPs because of emotional, ethical, religious and social issues involved in it. In our country, MTP was legalised in 1971 with some restrictions to avoid its misuse. But some people misuse the facility of MTP by first getting the sex of the foetus determined and then getting the abortion done if the foetus is female. Such female foeticide is illegal.

If MTP is misused for illegal female foeticides, then why MTPs ? The answer is to get rid of unwanted pregnancies either due to casual unprotected intercourse or failure of the contraceptive used during coitus or rapes. MTPs are also necessary where pregnancy can be harmful or even fatal to the mother or to the foetus or both. MTP is comparatively safe up to 12 weeks (the first trimester) of pregnancy. It is much more riskier, when the MTP (abortion) is performed in the second trimester.

Majority of the MTPs are done illegally by unqualified quacks which are not only unsafe but could be fatal also. The other dangerous trend is the amniocentesis to determine the sex of the foetus. If the foetus is found to be female, in many cases, it is followed by MTP. There must be a check on such MTPs, as these are dangerous both for the mother and the foetus. There must be effective counselling about unprotected coitus and the risk factors involved in illegal abortions. The mentioned unhealthy trend of illegal abortions could be reversed by providing more health care facilities.

SEXUALLY TRANSMITTED DISEASES (STDs)

The term 'Sexually Transmitted Diseases' (STDs) refers to *the diseases or infections which are transmitted through sexual intercourse*'. STDs are also called **Venereal Diseases (VD) or Reproductive Tract Infections (RTI).** They are caused by viruses, bacteria, chlamydia or protozoa. Gonorrhoea, syphilis, genital herpes, chlamydiasis, genital warts, trichomoniasis, hepatitis-B and HIV leading to AIDs are some common examples of STDs.

Transmission. Transmission of some of the STDs (such as hepatitis-B, HIV etc.) can also occur (*i*) by sharing of injection needles, surgical instruments etc., with infected persons (*ii*) by transfusion of blood or (*iii*) from an infected mother to foetus.

Symptoms. Many of the STDs remain unnoticed initially, because the early symptoms of these diseases are minor and include itching, fluid discharge, slight pain, swellings etc., in the genital region. Infected females may often be asymptomatic (without symptoms) and hence, may remain undetected for long. Absence or less significant symptoms in the early stages of infection and the social stigma attached to the STDs, deter infected persons from going for timely diagnosis and proper treatment of the disease.

Complications. STDs often lead to complications like **Pelvic Inflammatory Diseases (PID),** abortions (termination of pregnancy before the foetus reaches the stage of viability), **still birth** (birth of a dead foetus), **ectopic pregnancies** (implantation of the fertilised ovum outside of the uterine cavity resulting in a poorly developed decidual reaction in the uterus), **infertility** or even **cancer** of the reproductive tract.

STDs are a major threat to a healthy society. Except for hepatitis-B, genital herpes and HIV infections, other STDs are completely curable if detected early and treated properly. Therefore, prevention or early detection and cure of these diseases are given prime consideration under the reproductive healthcare programmes. Though, all persons are vulnerable to STD infections, but the incidence of these diseases is very high among persons in the age group of 15–24 years.

Prevention of STDs. The STD infections can be prevented by adhering to the following simple principles :

(*i*) Avoid sex with unknown partners/multiple partners.

(*ii*) Always use condoms during coitus.

(*iii*) If there is any doubt, one must consult a qualified doctor. If STD is detected one should get complete treatment.

Test for STDs. STDs can be confirmed through following tests.

(*i*) Culture and microscopic examination of the pathogen with specific staining.

(*ii*) Detection of antigen/antibody interaction using ELISA technique.

(*iii*) DNA hybridisation using tag molecules and Polymerase Chain Reaction (PCR) technique.

Some important STDs, their pathogen, symptoms, detection techniques and treatment are given in Table 10.1.

INFERTILITY

There are many couples which are infertile or are unable to produce children inspite of unprotected sexual activities. Thus, '*the inability to conceive or produce children even after 2 years of unprotected sexual cohabitation is called infertility*'. There are many reasons for infertility. These reasons may be physical, congenital, diseases, drugs, immunological or even psychological. Presently, there are specialised health care units (infertility clinics, etc.) in our country, which could help in diagnosis and corrective treatment of some of the infertility related disorders, and enable the childless couples to have children. The childless couples are assisted to have children through certain special techniques commonly called **Assisted Reproductive Technologies (ART).** Some important ARTs are described below.

1. *In Vitro* Fertilization (IVF). It invovles fertilization of ovum outside the body (in almost similar conditions as that in the body) followed by **Embryo Transfer (ET).** This technique is popularly known as 'test-tube baby' programme. In this technique, ova from the wife/donor female and sperms from the husband/donor male are collected and are induced to get fuse to form zygote under simulated conditions in the laboratory. The zygote is allowed to divide (undergo cleavage). The zygote or early embryos (with up to 8 blastomeres) is then transferred into the fallopian tube. It is called **Zygote Intra Fallopian Transfer (ZIFT).** If the embryo is with more than 8 blastomeres, it is transferred into the uterus. It is called **Intra Uterine Transfer (IUT).** The transferred embryos complete their further developed within the uterus. It is to be noted that in test-tube baby programme, the baby is not reared in the test-tube. The success rate of this technique of producing test-tube babies is less than 20%.

The embryo formed by ***in vitro* fertilization** *i.e.,* fusion of gametes within the female can be used for such transfer to assist such females who cannot conceive due to one reason or the other.

Table 10.1. Some Important STDs, their Pathogen, Symptoms, Detection Techniques and Treatments

Name of STD	*Causal Agent*	*Symptoms*	*Detection Techniques*	*Treatment*
1. Chlamydiasis	*Chlamydia trachomatis* (An obligate intracellular pathogen. Which affect both male and female partners).	Urethritis epididymitis (with unilateral scrotal pain, tenderness and swelling) yellowish mucus and pus, cervicitis, inflammation of fallopian tubes, rectal pain with mucus and occasional bleeding).	Detection technique include Gram–ve discharge, antigen detection, microscopic examination of scraping, cell culture and nucleic acid hybridization.	Course of antibiotic drugs like tetracycline/deoxycycline/erythromycin/azithromycin.
2. Gonorrhoea	*Neisseria gonorrhoeae* (A bacterium which spread through sexual contact, common toilets and under clothes).	The bacterium lives in genital tract, produces pus-containing discharge, pain around genitalia and burning sensation during urination. It may leads to arthritis and eye infection in children of gonorrhoea affected mothers.	Diagnosis is carried out by clinical symptoms, Gram staining discharge and culture.	Disease can be cured through the use of appropriate antibiotics like penicillin, ampicillin etc.
3. Syphilis	*Treponema pallidum* (A spirochaete bacterium) Transmission occurs through sexual contact and from mother to children.	Initially there is indurated infections and painless ulcer or cancer on the genitals and swelling of local lymph glands. In the second stage, cancer is healed and there are skin lesions, raches, hair loss, swollen joints and flu-like illness occasionally. In the tertiary stages chronic ulcers appear on palate, nose and lower leg. There can be paralysis, brain damage, blindness, heart trouble and aortic impairment.	The disease is diagnosed by clinical symptoms, microscopic examination and antibody detection.	The disease is curable if treated with appropriate antibiotics like penicillin, tetracycline, etc.

4. Trichomoniasis	*Trichomonas vaginalis* (A quadriflagellated protozoan. The parasite infects both males and females.	In females it causes vaginitis with foul smelling, yellow vaginal discharge and burning sensation. In males it causes urethritis, epididymitis and prostatitis resulting in pain and burning sensation.	The disease is diagnosed by clinical symptoms, microscopic examination, culture and immunofluorescent antibody staining.	Both partners be treated simultaneously with the standard drug metronidazole.
5. Genital Herpes	*Herpes simplex* (A virus). Transmitted sexually through genital secretions but also contact with viroids and genitalia.	Vesiculopustular lesions followed by clusters of painful erythemations ulcers over external genitalia and perianal regions. Symptoms are more severe in females. Infection of new born can occur in case of infected females. There is fever, headache, itching vaginal and urethral discharge, with tender inguinal lymphadenopathy (swelling of lymph nodes). Urethral and cervix infections also occur.	Diagnosis is carried out by clinical symptoms, antigen detection, PCR and nucleic acid hybridisation.	Treatment is carried out by acyclovir, valacyclovir or fameiclovir.
6. Genital Warts	*Human Palilloma Virus* (HPV)	Warts (benign hard outgrowths with horny surface) develop over the skin and mucosal surface of external genitalia and parianal area. In woman infection may enter vagina and cervix.	Diagnosis is made through clinical symptoms, antibody defection, culture and DNA hybridization.	Treatment requires special skill especially in case of internal warts, cryosurgery is used for the removal of warts. Podophyllum preparation and podofilox are useful. Imiquimod and interferon inducer is also beneficial.

7. Chancroid	*Heamophilus ducrei* (A bacterium).	Ulcer appears at the site of infection generally over external genitalia. It is painful and bleeds easily. Nearby lymph nodes swell up and become tender.	The disease is diagnosed by clinical symptoms, staining of discharge and cell culture.	The effective antibiotics are ceftriaxone, erythromycin, ciprofloxacin and trimethoprim-sulphamethoxazole.
8. Hepatitis-B	*Hepatitis-B virus* Transmission may be through sexual contact, blood transmission, saliva, tears, intravenous drug abuse, tatooing, ear and nose piercing, sharing of razors.	Cirrhosis and possibly cancer of liver.	PCR and DNA hybridization.	Vaccines produced through recombinant DNA technology (*e.g.*, Recombivax-HB) are used to prevent Hepatitis-B infection.
9. Acquired Immune Deficiency Syndrome (AIDS)	Human immunodeficiency virus (HIV) (A retrovirus).	The virus attack on helper-T cells which stimulate antibody production of B-cells of the immune system, resulting in the loss of natural defence against opportunistic infections (infections to which a normal person is expected to be resistant).	HIV infection is detected by ELISA test; the infection is confirmed by Western blot analysis.	There is no cure for AIDs. Use of certain drugs can prolong the life of AIDs patient. These drugs include, Zidovudine, AZT (3'-azidoz', 3-dideoxy-thymidine) and didanosine.

The world's first test-tube baby was **Louise Joy Brown,** was born on July 25, 1978 in Oldham, Lancashire, England. The Indian's first test-tube baby was **Kum Harsha.** She was born on August 6, 1986 at K.E.M. Hospital, Mumbai, with the help of Dr. Indra Hinduja.

2. Gamete Intra Fallopian Transfer (GIFT). In this technique, ovum is collected from a donor female and is transferred into the fallopian tube of another female who cannot produce the ovum, but can provide suitable environment for fertilization and further development of embryo. Such transfer of the ovum is called **'Gamete Intra Fallopian Transfer'** or **(GIFT).**

3. Intra-Cytoplasmic Sperm Injection (ICSI). In this technique a sperm is directly injected into the ovum under appropriate conditions in the laboratory.

4. Artificial Insemination (AI). It is a technique by which the infertility cases of male partner either due to inability to inseminate the female or due to very low sperm counts in the ejaculates, could be corrected. In artificial insemination, the semen collected either from the husband or a healthy donor is artificially introduced either into the vagina or into the uterus of the female. The introduction of the semen directly into the uterus is termed as **Intra-Uterine Insemination (IUI).**

Presently, many options are available to deal with the problems of infertility, but all these techniques require expensive instrumentation and handling by specialised professionals having high precision. Therefore, there are only few such centres in the country and hence their benefits are affordable to only a limited number of people. The adoption of these methods are also interfered by emotional, religious and social factors. The ultimate aim of these procedures is to have children. In country like ours, there are so many orphaned and destitute children. Adoption of such children, which is also permitted by our country's law, could be one of the best methods for couples looking for parenthood.

REVIEW EXERCISES

MULTIPLE CHOICE TYPE QUESTIONS

1. What is correct about amniocentesis ?
(*a*) Intra-uterine diagnosis
(*b*) Withdrawal of amniotic fluid from pregnant woman
(*c*) Chemical analysis of fluid of pregnant woman
(*d*) Culturing amniotic cells and study of metaphasic chromosomes to identify chromosomal abnormalities.

2. Tubectomy, a method of population control, is performed on
(*a*) Both males and females (*b*) Males only
(*c*) Female only (*d*) Only pregnant females.

3. World population day is
(*a*) 5th June (*b*) 11th July
(*c*) 4th October (*d*) 21st March.

4. Copper-T
(*a*) suppresses fertilisation by covering ovum
(*b*) acts as berrier
(*c*) prevents implantation of blastocyst
(*d*) prevents cleavage.

5. Family planning programme was introduced in
(*a*) 1920 (*b*) 1930
(*c*) 1950 (*d*) 1951.

6. Which one of the following related to males ?
(*a*) Oral pills (*b*) Vasectomy
(*c*) Tubectomy (*d*) Copper-T.

7. Which one of the following is the most widely accepted method of contraception in India as at present ?
(*a*) IUDs (*b*) Cervical caps
(*c*) Tubectomy (*d*) Diaphragms.

8. *In vitro* fertilisation is a technique that involves transfer of which one of the following into the fallopian tube ?
(*a*) Embryo of 32 cell stage
(*b*) Zygote only
(*c*) Embryo only upto 8 cell stage
(*d*) Either zygote or early embryo upto 8 cell stage.

9. Medical termination of pregnancy (MTP) is considered safe upto how many weeks of pregnancy ?
(*a*) Six weeks (*b*) Eight weeks
(*c*) Twelve weeks (*d*) Eighteen weeks.

10. Tubal ligation
(*a*) Cut in vasa deferentia (*b*) Cut in Fallopian tubes
(*c*) Tying fallopian tubes (*d*) Tying of vasa deferentia.

OBJECTIVE TYPE QUESTIONS

1. Fill up in the blanks with suitable word/s.
(*i*) Amniocentesis has been developed for detecting foetal abnormalities if any, by analysing the .... defect.
(*ii*) A rapid increase in human population over a relatively short period is called ....
(*iii*) According to the census report 2001, the India's population growth was around .... per cent.
(*iv*) In India, family planning programme was initiated in the year ....
(*v*) IUDs are ideal contraceptives for the females, who want to delay .... or ....

2. Match the term in Column I with those in Column II.

Column I	Column II
(*i*) Saheli	(*a*) Prevention of conception
(*ii*) Demography	(*b*) Sterilisation in female
(*iii*) Multiload-375	(*c*) Central Drug Research Institute, Lucknow
(*iv*) Condoms	(*d*) Scientific study of human population
(*v*) Tubectomy	(*e*) Suppression of sperm motality

3. Correct the following statements :
(*i*) Surgical methods of contraception prevent gamete formation.
(*ii*) All sexually transmitted diseases are completely curable.
(*iii*) Oral pills are very popular contraceptives among the rural women.
(*iv*) In ET techniques, embryos are always transferred into the uterus.

TEST QUESTIONS

1. What is reproductive health ?
2. In what way amniocentosis is being misused ?
3. What are the main reasons of population explosion ?
4. What are STDs ? What are the complication which may be caused in women by STDs ?
5. How does Cu-T act as an effective contraceptive for human females ?
6. How do copper and hormone releasing IUDs act as contraceptives ? Explain.
7. Mention the features of an ideal contraceptive.
8. Describe various surgical methods of birth control.
9. Describe natural methods of birth control.
10. Write an essay on the assisted reproductive technologies.